FREE

Free Study Tips DVD

In addition to the tips and content in this guide, we have created a FREE DVD with helpful study tips to further assist your exam preparation. **This FREE Study Tips DVD provides you with top-notch tips to conquer your exam and reach your goals.**

Our simple request in exchange for the strategy-packed DVD is that you email us your feedback about our study guide. We would love to hear what you thought about the guide, and we welcome any and all feedback—positive, negative, or neutral. It is our #1 goal to provide you with top-quality products and customer service.

To receive your **FREE Study Tips DVD**, email freedvd@apexprep.com. Please put "FREE DVD" in the subject line and put the following in the email:

a. The name of the study guide you purchased.

b. Your rating of the study guide on a scale of 1-5, with 5 being the highest score.

c. Any thoughts or feedback about your study guide.

d. Your first and last name and your mailing address, so we know where to send your free DVD!

Thank you!

TSI Math Study Guide
2021 & 2022
TSI Preparation with 3 Mathematics Practice Tests
[Book Updated for the New Texas 2.0 Outline]

Matthew Lanni

Written and edited by APEX Publishing.

ISBN 13: 9781637750797
ISBN 10: 163775079X

For additional information or for bulk orders, contact info@apexprep.com.

Table of Contents

Test Taking Strategies

1. Reading the Whole Question

A popular assumption in Western culture is the idea that we don't have enough time for anything. We speed while driving to work, we want to read an assignment for class as quickly as possible, or we want the line in the supermarket to dwindle faster. However, speeding through such events robs us from being able to thoroughly appreciate and understand what's happening around us. While taking a timed test, the feeling one might have while reading a question is to find the correct answer as quickly as possible. Although pace is important, don't let it deter you from reading the whole question. Test writers know how to subtly change a test question toward the end in various ways, such as adding a negative or changing focus. If the question has a passage, carefully read the whole passage as well before moving on to the questions. This will help you process the information in the passage rather than worrying about the questions you've just read and where to find them. A thorough understanding of the passage or question is an important way for test takers to be able to succeed on an exam.

2. Examining Every Answer Choice

Let's say we're at the market buying apples. The first apple we see on top of the heap may *look* like the best apple, but if we turn it over we can see bruising on the skin. We must examine several apples before deciding which apple is the best. Finding the correct answer choice is like finding the best apple. Although it's tempting to choose an answer that seems correct at first without reading the others, it's important to read each answer choice thoroughly before making a final decision on the answer. The aim of a test writer might be to get as close as possible to the correct answer, so watch out for subtle words that may indicate an answer is incorrect. Once the correct answer choice is selected, read the question again and the answer in response to make sure all your bases are covered.

3. Eliminating Wrong Answer Choices

Sometimes we become paralyzed when we are confronted with too many choices. Which frozen yogurt flavor is the tastiest? Which pair of shoes look the best with this outfit? What type of car will fill my needs as a consumer? If you are unsure of which answer would be the best to choose, it may help to use process of elimination. We use "filtering" all the time on sites such as eBay® or Craigslist® to eliminate the ads that are not right for us. We can do the same thing on an exam. Process of elimination is crossing out the answer choices we know for sure are wrong and leaving the ones that might be correct. It may help to cover up the incorrect answer choice. Covering incorrect choices is a psychological act that alleviates stress due to the brain being exposed to a smaller amount of information. Choosing between two answer choices is much easier than choosing between all of them, and you have a better chance of selecting the correct answer if you have less to focus on.

4. Sticking to the World of the Question

When we are attempting to answer questions, our minds will often wander away from the question and what it is asking. We begin to see answer choices that are true in the real world instead of true in the world of the question. It may be helpful to think of each test question as its own little world. This world may be different from ours. This world may know as a truth that the chicken came before the egg or may assert that two plus two equals five. Remember that, no matter what hypothetical nonsense may be in the question, assume it to be true. If the question states that the chicken came before the egg, then choose your answer based on that truth. Sticking to the world of the question means placing all of our biases and assumptions aside and relying on the question to guide us to the correct answer. If we are simply looking for answers that are correct based on our own judgment, then we may choose incorrectly. Remember an answer that is true does not necessarily answer the question.

5. Key Words

If you come across a complex test question that you have to read over and over again, try pulling out some key words from the question in order to understand what exactly it is asking. Key words may be words that surround the question, such as *main idea, analogous, parallel, resembles, structured,* or *defines.* The question may be asking for the main idea, or it may be asking you to define something. Deconstructing the sentence may also be helpful in making the question simpler before trying to answer it. This means taking the sentence apart and obtaining meaning in pieces, or separating the question from the foundation of the question. For example, let's look at this question:

> Given the author's description of the content of paleontology in the first paragraph, which of the following is most parallel to what it taught?

The question asks which one of the answers most *parallels* the following information: The *description* of paleontology in the first paragraph. The first step would be to see *how* paleontology is described in the first paragraph. Then, we would find an answer choice that parallels that description. The question seems complex at first, but after we deconstruct it, the answer becomes much more attainable.

6. Subtle Negatives

Negative words in question stems will be words such as *not, but, neither,* or *except.* Test writers often use these words in order to trick unsuspecting test takers into selecting the wrong answer—or, at least, to test their reading comprehension of the question. Many exams will feature the negative words in all caps (*which of the following is NOT an example*), but some questions will add the negative word seamlessly into the sentence. The following is an example of a subtle negative used in a question stem:

> According to the passage, which of the following is *not* considered to be an example of paleontology?

If we rush through the exam, we might skip that tiny word, *not,* inside the question, and choose an answer that is opposite of the correct choice. Again, it's important to read the question fully, and double check for any words that may negate the statement in any way.

7. Spotting the Hedges

The word "hedging" refers to language that remains vague or avoids absolute terminology. Absolute terminology consists of words like *always, never, all, every, just, only, none,* and *must.* Hedging refers to words like *seem, tend, might, most, some, sometimes, perhaps, possibly, probability,* and *often.* In some cases, we want to choose answer choices that use hedging and avoid answer choices that use absolute terminology. It's important to pay attention to what subject you are on and adjust your response accordingly.

8. Restating to Understand

Every now and then we come across questions that we don't understand. The language may be too complex, or the question is structured in a way that is meant to confuse the test taker. When you come across a question like this, it may be worth your time to rewrite or restate the question in your own words in order to understand it better. For example, let's look at the following complicated question:

> Which of the following words, if substituted for the word *parochial* in the first paragraph, would LEAST change the meaning of the sentence?

Let's restate the question in order to understand it better. We know that they want the word *parochial* replaced. We also know that this new word would "least" or "not" change the meaning of the sentence. Now let's try the sentence again:

> Which word could we replace with *parochial*, and it would not change the meaning?

Restating it this way, we see that the question is asking for a synonym. Now, let's restate the question so we can answer it better:

> Which word is a synonym for the word *parochial*?

Before we even look at the answer choices, we have a simpler, restated version of a complicated question.

9. Predicting the Answer

After you read the question, try predicting the answer *before* reading the answer choices. By formulating an answer in your mind, you will be less likely to be distracted by any wrong answer choices. Using predictions will also help you feel more confident in the answer choice you select. Once you've chosen your answer, go back and reread the question and answer choices to make sure you have the best fit. If you have no idea what the answer may be for a particular question, forego using this strategy.

10. Avoiding Patterns

One popular myth in grade school relating to standardized testing is that test writers will often put multiple-choice answers in patterns. A runoff example of this kind of thinking is that the most common answer choice is "C," with "B" following close behind. Or, some will advocate certain made-up word patterns that simply do not exist. Test writers do not arrange their correct answer choices in any kind of pattern; their choices are randomized. There may even be times where the correct answer choice will be the same letter for two or three questions in a row, but we have no way of knowing when or if this might happen. Instead of trying to figure out what choice the test writer probably set as being correct, focus on what the *best answer choice* would be out of the answers you are presented with. Use the tips above, general knowledge, and reading comprehension skills in order to best answer the question, rather than looking for patterns that do not exist.

FREE DVD OFFER

Achieving a high score on your exam depends not only on understanding the content, but also on understanding how to apply your knowledge and your command of test taking strategies. **Because your success is our primary goal, we offer a FREE Study Tips DVD. It provides top-notch test taking strategies to help you optimize your testing experience.**

Our simple request in exchange for the strategy-packed DVD is that you email us your feedback about our study guide.

To receive your **FREE Study Tips DVD**, email freedvd@apexprep.com. Please put "FREE DVD" in the subject line and put the following in the email:

a. The name of the study guide you purchased.

b. Your rating of the study guide on a scale of 1-5, with 5 being the highest score.

c. Any thoughts or feedback about your study guide.

d. Your first and last name and your mailing address, so we know where to send your free DVD!

Introduction

Function of the Test

The Texas Success Initiative Assessment (TSI Assessment) is an assessment required by Texas universities in order to determine if students are prepared for college-level course work. The full assessment includes Mathematics and English Language Arts and Reading sections. Exemptions for the exam are available depending on scores from the ACT, SAT, TAKS, or transfer work from another university. Those who perform well on the TSI may be enrolled into a college-level course. Those who do not perform well may be placed into a developmental course to ensure preparation for college-level coursework in the future.

The Mathematics section is one of the two sections on the exam. This guide is intended to help students focus on the necessary skills and knowledge assessed by the Mathematics section so that they can perform well and achieve a college-ready score.

Test Administration

For most Texas universities, the rules for taking the TSI assessment are similar. Students must first complete an application for admission into the university. Next, they will be assigned a TCC ID number. After, students must take a mandatory pre-assessment activity before they take the TSI. This activity is given at the same university where students are admitted to take the TSI exam. The pre-assessment activity includes an explanation of the importance of the TSI assessment, practice questions, explanation of developmental education, and information of educational resources on campus. Visit your university's website in order to find test dates and times for the pre-assessment activity as well as for the TSI assessment.

Again, those who do not meet the requirements of the TSI assessment in one or more subject areas may be placed in a developmental course or intervention to improve knowledge in that area. Note that the diagnostic test is used for students who score lower than the college readiness benchmark. Students who are not satisfied with their scores may retake the TSI at any time, but it is recommended that time be set aside for studying. Further information about retesting must be obtained through the university students wish to attend.

Test Format

The content areas of the full TSI assessment include Mathematics and English Language Arts and Reading. The table below details the four major content areas in the Mathematics sections and the topics covered within those areas:

Content Area	Topics Covered
Quantitative Reasoning	Ratios, proportions, and percents
	Rational and irrational numbers
	Manipulation of linear equations, inequalities, and expressions
Algebraic Reasoning	Solving linear equations, inequalities, and systems of linear equations
	Evaluation of linear functions
	Quadratic, polynomial, exponential, rational, and radical equations and expressions
Geometric and Spatial Reasoning	Connections between geometry and algebraic equations
	Transformations, congruence, similarity, and symmetry
	Unit conversion
	Perimeter, area, surface area, and volume
	Right triangle relationships and basic trigonometry
Probabilistic and Statistical Reasoning	Classifying and analyzing representations of data
	Measures of center and spread of data
	Probability

The Mathematics section contains 20 multiple-choice questions. Calculators are not permitted. The exam is not timed, so test takers can take as long as they need. Most students take three to five hours to complete the entire exam.

Scoring

Students will have immediate access to their scores once they finish the TSI assessment. Their performance on the math section (and the other section) helps schools determine whether the student will enter a college-level math course (and which course is appropriate for them) or a college readiness course or math intervention to help the student develop skills for college-level math courses. A college-ready score in Mathematics involves a passing score on the standard test or an EFL of 6 on the diagnostic test.

Quantitative Reasoning

Linear Equations, Inequalities, and Systems

Linear Equations

An **equation in one variable** is a mathematical statement where two algebraic expressions in one variable, usually x, are set equal. To solve the equation, the variable must be isolated on one side of the equals sign. The addition and multiplication principles of equality are used to isolate the variable. The **addition principle of equality** states that the same number can be added to or subtracted from both sides of an equation. Because the same value is being used on both sides of the equals sign, equality is maintained. For example, the equation $2x - 3 = 5x$ is equivalent to both:

$$2x - 3 + 2 = 5x + 2$$

and

$$2x - 3 - 5 = 5x - 5$$

This principle can be used to solve the following equation: $x + 5 = 4$. The variable x must be isolated, so to move the 5 from the left side, subtract 5 from both sides of the equals sign. Therefore:

$$x + 5 - 5 = 4 - 5$$

So, the solution is $x = -1$. This process illustrates the idea of an **additive inverse** because subtracting 5 is the same as adding -5. Basically, add the opposite of the number that must be removed to both sides of the equals sign. The multiplication principle of equality states that equality is maintained when both sides of an equation are multiplied or divided by the same number. For example, $4x = 5$ is equivalent to both $16x = 20$ and $x = \frac{5}{4}$. Multiplying both sides times 4 and dividing both sides by 4 maintains equality. Solving the equation $6x - 18 = 5$ requires the use of both principles. First, apply the addition principle to add 18 to both sides of the equals sign, which results in $6x = 23$. Then use the multiplication principle to divide both sides by 6, giving the solution $x = \frac{23}{6}$. Using the multiplication principle in the solving process is the same as involving a multiplicative inverse. A **multiplicative inverse** is a value that, when multiplied by a given number, results in 1. Dividing by 6 is the same as multiplying by $\frac{1}{6}$, which is both the reciprocal and multiplicative inverse of 6.

When solving linear equations, check the answer by plugging the solution back into the original equation. If the result is a false statement, something was done incorrectly during the solution procedure. Checking the example above gives the following:

$$6 \times \frac{23}{6} - 18 = 23 - 18 = 5$$

Therefore, the solution is correct.

Some equations in one variable involve fractions or the use of the distributive property. In either case, the goal is to obtain only one variable term and then use the addition and multiplication principles to isolate that variable. Consider the equation $\frac{2}{3}x = 6$. To solve for x, multiply each side of the equation by the reciprocal of $\frac{2}{3}$, which is $\frac{3}{2}$. This step results in $\frac{3}{2} \times \frac{2}{3}x = \frac{3}{2} \times 6$, which simplifies into the solution $x = 9$. Now consider the equation:

$$3(x + 2) - 5x = 4x + 1$$

Use the distributive property to clear the parentheses. Therefore, multiply each term inside the parentheses by 3. This step results in:

$$3x + 6 - 5x = 4x + 1$$

Next, collect like terms on the left-hand side. **Like terms** are terms with the same variable or variables raised to the same exponent(s). Only like terms can be combined through addition or subtraction. After collecting like terms, the equation is:

$$-2x + 6 = 4x + 1$$

Finally, apply the addition and multiplication principles. Add $2x$ to both sides to obtain:

$$6 = 6x + 1$$

Then, subtract 1 from both sides to obtain $5 = 6x$. Finally, divide both sides by 6 to obtain the solution $\frac{5}{6} = x$.

Two other types of solutions can be obtained when solving an equation in one variable. There could be no solution, or the solution set could contain all real numbers. Consider the equation:

$$4x = 6x + 5 - 2x$$

First, the like terms can be combined on the right to obtain:

$$4x = 4x + 5$$

Next, subtract $4x$ from both sides. This step results in the false statement $0 = 5$. There is no value that can be plugged into x that will ever make this equation true. Therefore, there is no solution. The solution procedure contained correct steps, but the result of a false statement means that no value satisfies the equation. The symbolic way to denote that no solution exists is $\emptyset$.

Next, consider the equation:

$$5x + 4 + 2x = 9 + 7x - 5$$

Combining the like terms on both sides results in:

$$7x + 4 = 7x + 4$$

The left-hand side is exactly the same as the right-hand side. Using the addition principle to move terms, the result is $0 = 0$, which is always true. Therefore, the original equation is true for any number, and the solution set is all real numbers. The symbolic way to denote such a solution set is $\mathbb{R}$, or in interval notation, $(-\infty, \infty)$.

The Connection Between Proportional Relationships and Linear Equations

Linear growth involves a quantity, the **dependent variable**, increasing or decreasing at a constant rate as another quantity, the **independent variable**, increases as well. The graph of linear growth is a straight line. Linear growth is represented as the following equation: $y = mx + b$, where *m* is the **slope** of the line, also known as the **rate of change**, and *b* is the ***y*-intercept**. If the *y*-intercept is 0, then the linear growth is actually known as **direct variation**. If the slope is positive, the dependent variable increases as the independent variable increases, and if the slope is negative, the dependent variable decreases as the independent variable increases.

A linear function that models a linear relationship between two quantities is of the form $y = mx + b$, or in function form $f(x) = mx + b$. In a linear function, the value of *y* depends on the value of *x*, and *y* increases or decreases at a constant rate as *x* increases. Therefore, the independent variable is *x*, and the dependent variable is *y*. The graph of a linear function is a line, and the constant rate can be seen by looking at the steepness, or **slope**, of the line. If the line increases from left to right, the slope is positive. If the line slopes downward from left to right, the slope is negative. In the function, *m* represents slope. Each point on the line is an **ordered pair** (*x*, *y*), where *x* represents the *x*-coordinate of the point and *y* represents the *y*-coordinate of the point. The point where *x* = 0 is known as the *y*-intercept, and it is the place where the line crosses the *y*-axis. If *x* = 0 is plugged into $f(x) = mx + b$, the result is $f(0) = b$, so therefore, the point (0, *b*) is the *y*-intercept of the line. The derivative of a linear function is its slope.

Consider the following situation. A taxicab driver charges a flat fee of $2 per ride and $3 a mile. This statement can be modeled by the function $f(x) = 3x + 2$ where x represents the number of miles and $f(x) = y$ represents the total cost of the ride. The total cost increases at a constant rate of $2 per mile, and that is why this situation is a linear relationship. The slope $m = 3$ is equivalent to this rate of change. The flat fee of $2 is the y-intercept. It is the place where the graph crosses the x-axis, and it represents the cost when $x = 0$, or when no miles have been traveled in the cab. The y-intercept in this situation represents the flat fee.

Linear Inequalities

A **linear equation in x** can be written in the form $ax + b = 0$. A **linear inequality** is very similar, although the equals sign is replaced by an inequality symbol such as $<, >, \leq$, or $\geq$. In any case, *a* can never be 0. Some examples of linear inequalities in one variable are:

$$2x + 3 < 0$$

and

$$4x - 2 \leq 0$$

Solving an inequality involves finding the set of numbers that when plugged into the variable, make the inequality a true statement. These numbers are known as the **solution set** of the inequality. To solve an inequality, use the same properties that are necessary in solving equations. First, add or subtract variable terms and/or constants to obtain all variable terms on one side of the equals sign and all constant terms on the other side. Then, either multiply or divide both sides by the same number to obtain an inequality that gives the solution set. When multiplying or dividing by a negative number, change the direction of the inequality symbol. The solution set can be graphed on a number line. Consider the linear inequality:

$$-2x - 5 > x + 6$$

First, add 5 to both sides and subtract x from both sides to obtain $-3x > 11$. Then, divide both sides by -3, making sure to change the direction of the inequality symbol. These steps result in the solution $x < -\frac{11}{3}$. Therefore, any number less than $-\frac{11}{3}$ satisfies this inequality.

Algebraically Solving Linear Equations or Inequalities in One Variable

A **linear equation in one variable** can be solved using the following steps:

1. Simplify both sides of the equation by removing all parentheses, using the distributive property, and collecting all like terms.

2. Collect all variable terms on one side of the equation and all constant terms on the other side by adding the same quantity to or subtracting the same quantity from both sides.

3. Isolate the variable by either multiplying or dividing both sides of the equation by the same number.

4. Check the answer.

The only difference between solving linear inequalities versus equations is that when multiplying by a negative number or dividing by a negative number, the direction of the inequality symbol must be reversed.

If an equation contains multiple fractions, it might make sense to clear the equation of fractions first by multiplying all terms by the least common denominator. Also, if an equation contains several decimals, it might make sense to clear the decimals as well by multiplying times a factor of 10. If the equation has decimals in the hundredths place, multiply every term in the equation by 100.

Systems of Linear Equations and Equalities

Systems of Two Linear Equations in Two Variables

An example of a system of two linear equations in two variables is the following:

$$2x + 5y = 8$$

$$5x + 48y = 9$$

A solution to a system of two linear equations is an ordered pair that satisfies both the equations in the system. A system can have one solution, no solution, or infinitely many solutions. The solution can be found through a graphing technique. The solution to a system of equations is actually equal to the point where both lines intersects. If the lines intersect at one point, there is one solution and the system is said to be **consistent**. However, if the two lines are parallel, they will never intersect and there is no solution. In this case, the system is said to be **inconsistent.** Third, if the two lines are actually the same line, there are infinitely many solutions and the solution set is equal to the entire line. The lines are dependent.

Here is a summary of the three cases:

Consistent	Inconsistent	Dependent
One solution	No solution Ø	Infinite number of solutions ℝ
Lines intersect	Lines are parallel	Coincide/Same line
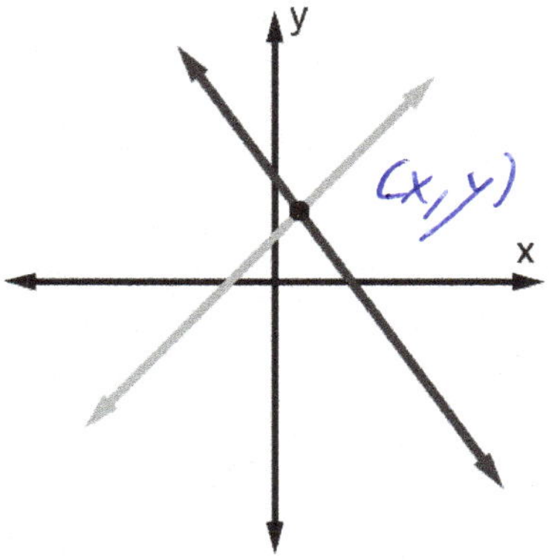	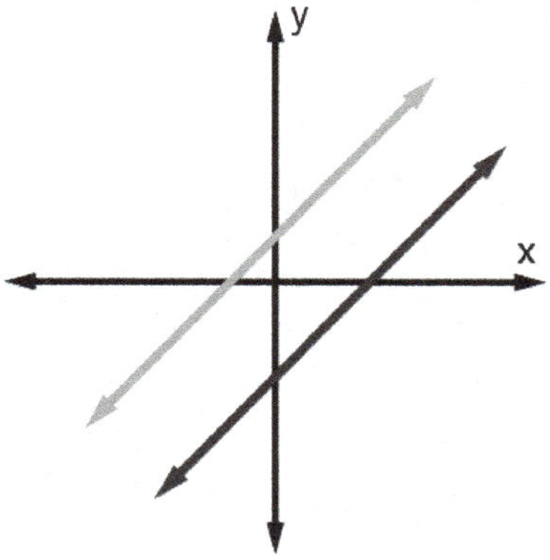	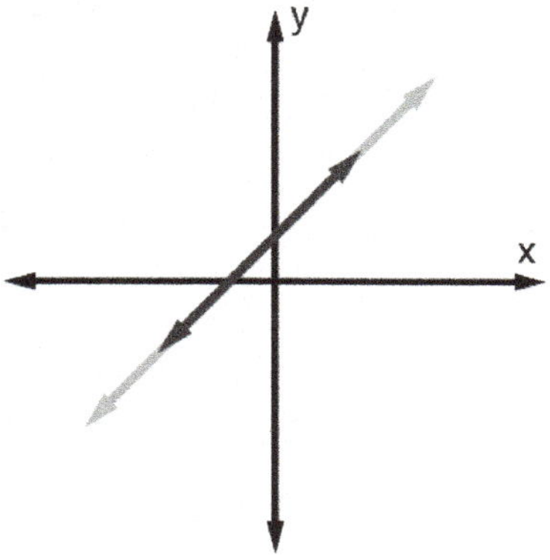

Consider the following system of equations:

$$y + x = 3$$

$$y - x = 1$$

To find the solution graphically, graph both lines on the same *xy*-plane. Graph each line using either a table of ordered pairs, the *x*- and *y*-intercepts, or slope and the *y*-intercept. Then, locate the point of intersection.

The graph is shown here:

The System of Equations $\begin{cases} y + x = 3 \\ y - x = 1 \end{cases}$

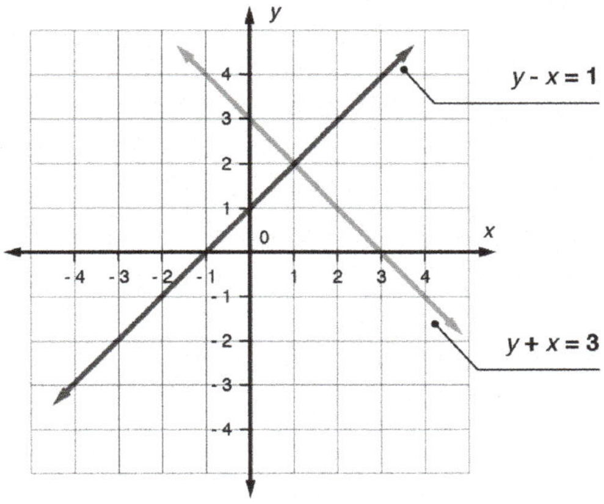

It can be seen that the point of intersection is the ordered pair (1, 2). This solution can be checked by plugging it back into both original equations to make sure it results in true statements. This process results in:

$$2 + 1 = 3$$

$$2 - 1 = 1$$

Both equations are true, so the solution is correct.

The following system has no solution:

$$y = 4x + 1$$

$$y = 4x - 1$$

Both lines have the same slope and different y-intercepts, so they are parallel, meaning that they run alongside each other and never intersect.

Finally, the following solution has infinitely many solutions:

$$2x - 7y = 12$$

$$4x - 14y = 24$$

Note that the second equation is equal to the first equation times 2. Therefore, they are the same line. The solution set can be written in set notation as $\{(x, y)|2x - 7y = 12\}$, which represents the entire line.

There are two algebraic methods to finding solutions to systems of two linear equations in two variables. The first is substitution. This process is better suited for systems when one of the equations is already solved for one variable, or when solving for one variable is easy to do. The equation that is already solved for is substituted into the other equation for that variable, and this process results in a linear equation in one variable. This equation can be solved for the given variable, and then that solution can be plugged into one of the original equations, which can then be solved for the other variable. This last step is known as **back-substitution**, and the end result is an ordered pair.

The following is an example of a system that is suited for substitution:

$$y = 4x + 2$$

$$2x + 3y = 9$$

The other method is known as **elimination,** or the **addition method**. This is better suited when the equations are in standard form $Ax + By = C$. The goal in this method is to multiply one or both equations times numbers that result in opposite coefficients. Then, add the equations together to obtain an equation in one variable. Solve for the given variable, then take that value and back-substitute to obtain the other part of the ordered pair solution.

The following is an example of a system that is suited for elimination:

$$2x + 3y = 8$$

$$4x - 2y = 10$$

Note that in order to check an answer when solving a system of equations, the solution must be checked in both original equations to show that it solves not only one of the equations, but both of them.

If either solution results in an untrue statement when inserted into the original equation, then there is no solution to the system. Finally, if throughout either solution procedure the process results in the variables dropping out, which gives a statement that is always true, there are infinitely many solutions.

Systems of Linear Inequalities in Two Variables

A system of linear inequalities in two variables consists of two inequalities in two variables, *x* and *y*. For example, the following is a system of linear inequalities in two variables:

$$\begin{cases} 4x + 2y < 1 \\ 2x - y \leq 0 \end{cases}$$

The curly brace on the left side shows that the two inequalities are grouped together. A solution of a single inequality in two variables is an ordered pair that satisfies the inequality. For example, (1, 3) is a solution of the linear inequality $y \geq x + 1$ because when plugged in, it results in a true statement. The graph of an inequality in two variables consists of all ordered pairs that make the solution true. Therefore, the entire solution set of a single inequality contains many ordered pairs, and the set can be graphed by using a half plane. A **half plane** consists of the set of all points on one side of a line. If the inequality consists of $>$ or $<$, the line is dashed because no solutions actually exist on the line shown. If the inequality consists of $\geq$ or $\leq$, the line is solid and solutions are on the line shown. To graph a linear inequality, graph the corresponding equation found by replacing the inequality symbol with an equals sign. Then pick a test point that exists on either side of the line. If that point results in a true statement when plugged into the original inequality, shade in the side containing the test point. If it results in a false statement, shade in the opposite side.

Solving a system of linear inequalities must be done graphically. Follow the process as described above for both given inequalities. The solution set to the entire system is the region that is in common to every graph in the system.

For example, here is the solution to the following system:

$$\begin{cases} y \geq 3 - x \\ y \leq -3 - x \end{cases}$$

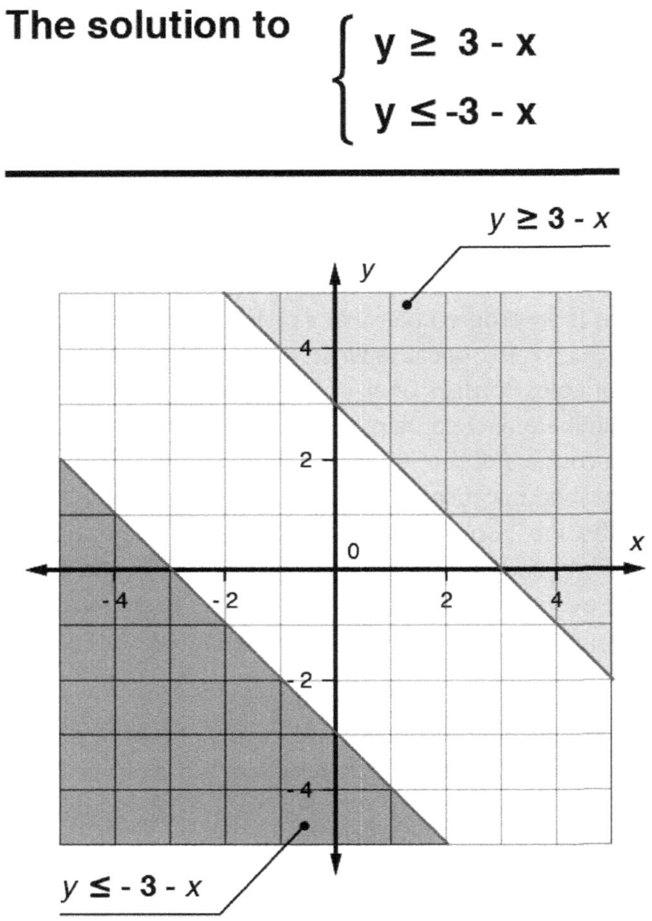

Note that there is no region in common, so this system has no solution.

Algebraic Expressions and Equations

An **algebraic expression** is a mathematical phrase that may contain numbers, variables, and mathematical operations. An expression represents a single quantity. For example, $3x + 2$ is an algebraic expression.

An **algebraic equation** is a mathematical sentence with two expressions that are equal to each other. That is, an equation must contain an equals sign, as in:

$$3x + 2 = 17$$

This statement says that the value of the expression on the left side of the equals sign is equivalent to the value of the expression on the right side. In an expression, there are not two sides because there is no equals sign. The equals sign (=) is the difference between an expression and an equation.

To distinguish an expression from an equation, just look for the equals sign.

Example: Determine whether each of these is an expression or an equation.

- $16 + 4x = 9x - 7$ Solution: Equation
- $-27x - 42 + 19y$ Solution: Expression
- $4 = x + 3$ Solution: Equation

Adding and Subtracting Linear Algebraic Expressions

To add and subtract linear algebraic expressions, you must combine like terms. Like terms are terms that have the same variable with the same exponent. In the following example, the x-terms can be added because the variable and exponent are the same. These terms add to be $9x$. Terms without a variable component are called constants. These terms will add to be nine.

Example: Add $(3x - 5) + (6x + 14)$

$3x - 5 + 6x + 14$	Rewrite without parentheses
$3x + 6x - 5 + 14$	Commutative property of addition
$9x + 9$	Combine like terms

When subtracting linear expressions, be careful to add the opposite when combining like terms. Do this by distributing -1, which is multiplying each term inside the second parenthesis by negative one. Remember that distributing -1 changes the sign of each term.

Example: Subtract $(17x + 3) - (27x - 8)$

$17x + 3 - 27x + 8$	Distributive Property
$17x - 27x + 3 + 8$	Commutative property of addition
$-10x + 11$	Combine like terms

Example: Simplify by adding or subtracting:

$(6m + 28z - 9) + (14m + 13) - (-4z + 8m + 12)$

$6m + 28z - 9 + 14m + 13 + 4z - 8m - 12$	Distributive Property
$6m + 14m - 8m + 28z + 4z - 9 + 13 - 12$	Commutative Property of Addition
$12m + 32z - 8$	Combine like terms

Using the Distributive Property to Generate Equivalent Linear Algebraic Expressions

The Distributive Property: $a(b + c) = ab + ac$

The **distributive property** is a way of taking a factor and multiplying it through a given expression in parentheses. Each term inside the parentheses is multiplied by the outside factor, eliminating the parentheses. The following example shows how to distribute the number 3 to all the terms inside the parentheses.

Example: Use the distributive property to write an equivalent algebraic expression:

$3(2x + 7y + 6)$	
$3(2x) + 3(7y) + 3(6)$	Distributive property
$6x + 21y + 18$	Simplify

Because $a - b$ can be written $a + (-b)$, the distributive property can be applied in the example below.

Example: Use the distributive property to write an equivalent algebraic expression.

$7(5m - 8)$	
$7[5m + (-8)]$	Rewrite subtraction as addition of -8
$7(5m) + 7(-8)$	Distributive property
$35m - 56$	Simplify

In the following example, note that the factor of 2 is written to the right of the parentheses but is still distributed as before.

Example: Use the distributive property to write an equivalent algebraic expression:

$(3m + 4x - 10)2$	
$(3m)2 + (4x)2 + (-10)2$	Distributive property
$6m + 8x - 20$	Simplify

Example: $-(-2m + 6x)$

In this example, the negative sign in front of the parentheses can be interpreted as $-1(-2m + 6x)$

$-1(-2m + 6x)$	
$-1(-2m) + (-1)(6x)$	Distributive property
$2m - 6x$	Simplify

Evaluating Simple Algebraic Expressions for Given Values of Variables

To evaluate an algebra expression for a given value of a variable, replace the variable with the given value. Then perform the given operations to simplify the expression.

Example: Evaluate $12 + x$ for $x = 9$

$12 + (9)$	Replace x with the value of 9 as given in the problem. It is a good idea to always use parentheses when substituting this value. This will be particularly important in the following examples.
21	Add

Now see that when x is 9, the value of the given expression is 21.

Example: Evaluate $4x + 7$ for $x = 3$

$4(3) + 7$	Replace the x in the expression with 3
$12 + 7$	Multiply (remember order of operations)
19	Add

Therefore, when x is 3, the value of the given expression is 19.

Example: Evaluate $-7m - 3r - 18$ for $m = 2$ and $r = -1$

$-7(2) - 3(-1) - 18$	Replace m with 2 and r with -1
$-14 + 3 - 18$	Multiply
-29	Add

So, when m is 2 and r is -1, the value of the given expression is -29.

Using Mathematical Terms to Identify Parts of Expressions and Describe Expressions

A **variable** is a symbol used to represent a number. Letters, like x, y, and z, are often used as variables in algebra.

A **constant** is a number that cannot change its value. For example, 18 is a constant.

A **term** is a constant, variable, or the product of constants and variables. In an expression, terms are separated by + and – signs. Examples of terms are $24x$, -32, and $15xyz$.

Like terms are terms that contain the same variables. For example, $6z$ and $-8z$ are like terms, and $9xy$ and $17xy$ are like terms. Constants, like 23 and 51, are like terms as well.

A **factor** is something that is multiplied by something else. A factor may be a constant, a variable, or a sum of constants or variables.

A **coefficient** is the numerical factor in a term that has a variable. In the term $16x$, the coefficient is 16.

Example: Given the expression, $6x - 12y + 18$, answer the following questions.

- How many terms are in the expression?
- Solution: 3
- Name the terms.
- Solution: 6x, –12y, and 18 (Notice that the minus sign preceding the 12 is interpreted to represent negative 12)
- Name the factors.
- Solution: 6, x, –12, y
- What are the coefficients in this expression?
- Solution: 6 and –12
- What is the constant in this expression?
- Solution: 18

Simplifying Rational Algebraic Expressions
When given a problem, it is necessary to determine the best form of an expression or equation to use, given the context. Usually this involves some algebraic manipulation. If an equation is given, the simplest form of the equation is best. Simplifying involves using the distributive property, collecting like terms, etc. If an equation is needed to be solved, properties involving performing the same operation on both sides of the equation must be used. For example, if a number is added to one side of the equals sign, it must be added to the other side as well. This maintains a true equation.

If an expression is given, simplifying can only involve properties allowing to rewrite the expression as an equivalent form. If there is no equals sign, mathematical operations cannot be performed on the expression, unless it is a rational expression. A **rational expression** can be written in the form of a fraction, in which the numerator and denominator are both polynomials and the denominator is not equal to zero. Rational expressions can always be multiplied by a form of 1. For example, consider the following rational expression involving radicals: $\frac{2}{\sqrt{2}}$. It is incorrect to write a fraction with a root in the denominator, and therefore the expression must be rationalized. Multiply the fraction times $\frac{\sqrt{2}}{\sqrt{2}}$, a form of 1. This results in:

$$\frac{2}{\sqrt{2}} \times \frac{\sqrt{2}}{\sqrt{2}} = \frac{2\sqrt{2}}{\sqrt{4}} = \frac{2\sqrt{2}}{2} = \sqrt{2}$$

which is the most suitable form of the expression.

Creating an Equivalent Form of an Algebraic Expression
Two algebraic expressions are equivalent if they represent the same value, even if they look different. To obtain an equivalent form of an algebraic expression, follow the laws of algebra. For instance, addition and multiplication are both commutative and associative. Therefore, terms in an algebraic expression can be added in any order and multiplied in any order. For example, $4x + 2y$ is equivalent to $2y + 4x$ and $y \times 2 + x \times 4$. Also, the distributive law allows a number to be distributed throughout parentheses, as in the following:

$$a(b + c) = ab + ac$$

The expressions on both sides of the equals sign are equivalent. Collecting like terms is also important when working with equivalent forms because the simplest version of an expression is always the easiest one to work with.

Note that an expression is not an equation, and therefore expressions cannot be multiplied times numbers, divided by numbers, or have numbers added to them or subtracted from them and still have equivalent expressions. These processes can only happen in equations when the same step is performed on both sides of the equals sign.

Factoring Expressions
A factorization of an algebraic expression can be found. Throughout the process, a more complicated expression can be decomposed into products of simpler expressions. To factor a polynomial, first determine if there is a greatest common factor. If there is, factor it out. For example, $2x^2 + 8x$ has a greatest common factor of $2x$ and can be written as $2x(x + 4)$. After this, determine if the remaining polynomial follows a factoring pattern. If the polynomial has two terms, it could be a difference of squares, a sum of cubes, or a difference of cubes.

If it falls into one of these categories, use the following rules:

$$a^2 - b^2 = (a + b)(a - b)$$

$$a^3 + b^3 = (a + b)(a^2 - ab + b^2)$$

$$a^3 - b^3 = (a - b)(a^2 + ab + b^2)$$

If there are three terms, and if the trinomial is a perfect square trinomial, it can be factored into the following:

$$a^2 + 2ab + b^2 = (a + b)^2$$

$$a^2 - 2ab + b^2 = (a - b)^2$$

If not, try factoring into a product of two binomials in the form of $(x + p)(x + q)$. For example, to factor $x^2 + 6x + 8$, determine what two numbers have a product of 8 and a sum of 6. Those numbers are 4 and 2, so the trinomial factors into $(x + 2)(x + 4)$.

Finally, if there are four terms, try factoring by grouping. First, group terms together that have a common monomial factor. Then, factor out the common monomial factor from the first two terms. Next, look to see if a common factor can be factored out of the second set of two terms that results in a common binomial factor. Finally, factor out the common binomial factor of each expression, for example:

$$xy - x + 5y - 5$$

$$x(y - 1) + 5(y - 1)$$

$$(y - 1)(x + 5)$$

After the expression is completely factored, check the factorization by multiplying it out; if this results in the original expression, then the factoring is correct. Factorizations are helpful in solving equations that consist of a polynomial set equal to 0. If the product of two algebraic expressions equals 0, then at least one of the factors is equal to 0. Therefore, factor the polynomial within the equation, set each factor equal to 0, and solve. For example, $x^2 + 7x - 18 = 0$ can be solved by factoring into:

$$(x + 9)(x - 2) = 0$$

Set each factor equal to 0, and solve to obtain $x = -9$ and $x = 2$.

Word Problems and Applications

Translating Phrases and Sentences into Expressions, Equations, and Inequalities

When presented with a real-world problem, the first step is to determine what unknown quantity must be solved for. Use a **variable**, such as x or t, to represent that unknown quantity. Sometimes there can be two or more unknown quantities. In this case, either choose an additional variable, or if a relationship exists between the unknown quantities, express the other quantities in terms of the original variable. After choosing the variables, form algebraic expressions and/or equations that represent the verbal statement in the problem.

The following table shows examples of vocabulary used to represent the different operations:

Addition	Sum, plus, total, increase, more than, combined, in all
Subtraction	Difference, less than, subtract, reduce, decrease, fewer, remain
Multiplication	Product, multiply, times, part of, twice, triple
Division	Quotient, divide, split, each, equal parts, per, average, shared

The combination of operations and variables form both mathematical expression and equations. The differences between expressions and equations are that there is no equals sign in an expression, and that expressions are evaluated to find an unknown quantity, while equations are solved to find an unknown quantity. Also, inequalities can exist within verbal mathematical statements. Instead of a statement of equality, expressions state quantities are *less than, less than or equal to, greater than*, or *greater than or equal to.* Another type of inequality is when a quantity is said to be not equal to another quantity ($\neq$).

The steps for solving inequalities in one variable are the same steps for solving equations in one variable. The addition and multiplication principles are used. However, to maintain a true statement when using the $<, \leq, >$, and $\geq$ symbols, if a negative number is either multiplied times both sides of an inequality or divided from both sides of an inequality, the sign must be flipped. For instance, consider the following inequality:

$$3 - 5x \leq 8$$

First, 3 is subtracted from each side to obtain $-5x \leq 5$. Then, both sides are divided by -5, while flipping the sign, to obtain $x \geq -1$. Therefore, any real number greater than or equal to -1 satisfies the original inequality.

Solving Real-World One- or Multi-Step Problems with Rational Numbers

One-step problems take only one mathematical step to solve. For example, solving the equation $5x = 45$ is a one-step problem because the one step of dividing both sides of the equation by 5 is the only step necessary to obtain the solution $x = 9$. The multiplication principle of equality is the one step used to isolate the variable. The equation is of the form $ax = b$, where a and b are rational numbers. Similarly, the addition principle of equality could be the one step needed to solve a problem. In this case, the equation would be of the form $x + a = b$ or $x - a = b$, for real numbers a and b.

A multi-step problem involves more than one step to find the solution, or it could consist of solving more than one equation. An equation that involves both the addition principle and the multiplication principle is a two-step problem, and an example of such an equation is $2x - 4 = 5$. To solve, add 4 to both sides and then divide both sides by 2. An example of a two-step problem involving two separate equations is $y = 3x, 2x + y = 4$. The two equations form a system that must be solved together in two variables. The system can be solved by the substitution method. Since y is already solved for in terms of x, replace y with $3x$ in the equation $2x + y = 4$, resulting in $2x + 3x = 4$. Therefore, $5x = 4$ and $x = \frac{4}{5}$. Because there are two variables, the solution consists of a value for both x and for y. Substitute $x = \frac{4}{5}$ into either original equation to find y. The easiest choice is $y = 3x$. Therefore:

$$y = 3 \times \frac{4}{5} = \frac{12}{5}$$

The solution can be written as the ordered pair $\left(\frac{4}{5}, \frac{12}{5}\right)$.

Real-world problems can be translated into both one-step and multi-step problems. In either case, the word problem must be translated from the verbal form into mathematical expressions and equations that can be solved using algebra. An example of a one-step real-world problem is the following: A cat weighs half as much as a dog living in the same house. If the dog weighs 14.5 pounds, how much does the cat weigh? To solve this problem, an equation can be used. In any word problem, the first step must be defining variables that represent the unknown quantities. For this problem, let x be equal to the unknown weight of the cat. Because two times the weight of the cat equals 14.5 pounds, the equation to be solved is: $2x = 14.5$. Use the multiplication principle to divide both sides by 2. Therefore, $x = 7.25$, and the cat weighs 7.25 pounds.

Most of the time, real-world problems require multiple steps. The following is an example of a multi-step problem: The sum of two consecutive page numbers is equal to 437. What are those page numbers? First, define the unknown quantities. If x is equal to the first page number, then $x + 1$ is equal to the next page number because they are consecutive integers. Their sum is equal to 437. Putting this information together results in the equation:

$$x + x + 1 = 437$$

To solve, first collect like terms to obtain

$$2x + 1 = 437$$

Then, subtract 1 from both sides and then divide by 2. The solution to the equation is $x = 218$. Therefore, the two consecutive page numbers that satisfy the problem are 218 and 219. It is always important to make sure that answers to real-world problems make sense. For instance, it should be a red flag if the solution to this same problem resulted in decimals, which would indicate the need to check the work. Page numbers are whole numbers; therefore, if decimals are found to be answers, the solution process should be double-checked for mistakes.

To solve problems, follow these steps: Identify the variables that are known, decide which equation should be used, substitute the numbers, and solve. To solve an equation for the amount of time that has elapsed since an event, use the equation T = L – E where T represents the elapsed time, L represents the later time, and E represents the earlier time. For example, the Minnesota Vikings have not appeared in the Super Bowl since 1976. If the year is now 2017, how long has it been since the Vikings were in the Super Bowl? The later time, L, is 2017, E = 1976 and the unknown is T. Substituting these numbers, the equation is T = 2017 – 1976, and so T = 41. It has been 41 years since the Vikings have appeared in the Super Bowl.

Questions involving total cost can be solved using the formula, C = I + T where C represents the total cost, I represents the cost of the item purchased, and T represents the tax amount. To find the length of a rectangle given the area = 32 square inches and width = 8 inches, the formula A = L × W can be used. Substitute 32 for A and substitute 8 for w, giving the equation 32 = L×8. This equation is solved by dividing both sides by 8 to find that the length of the rectangle is 4. The formula for volume of a rectangular prism is given by the equation V = L × W × H. If the length of a rectangular juice box is 4 centimeters, the width is 2 centimeters, and the height is 8 centimeters, what is the volume of this box? Substituting in the formula we find V = 4 × 2 × 8, so the volume is 64 cubic centimeters. In a similar fashion as those previously shown, the mass of an object can be calculated given the formula, Mass = Density × Volume.

Solving Real-World Problems Involving Percentages

Percentages are defined to be parts per one hundred. To convert a decimal to a percentage, move the decimal point two units to the right and place the percent sign after the number. Percentages appear in many scenarios in the real world. It is important to make sure the statement containing the percentage is translated to a correct mathematical expression. Be aware that it is extremely common to make a mistake when working with percentages within word problems.

An example of a word problem containing a percentage is the following: 35% of people speed when driving to work. In a group of 5,600 commuters, how many would be expected to speed on the way to their place of employment? The answer to this problem is found by finding 35% of 5,600. First, change the percentage to the decimal 0.35. Then compute the product:

$$0.35 \times 5{,}600 = 1{,}960$$

Therefore, it would be expected that 1,960 of those commuters would speed on their way to work based on the data given. In this situation, the word "of" signals to use multiplication to find the answer. Another way percentages are used is in the following problem: Teachers work 8 months out of the year. What percent of the year do they work? To answer this problem, find what percent of 12 the number 8 is, because there are 12 months in a year. Therefore, divide 8 by 12, and convert that number to a percentage:

$$\frac{8}{12} = \frac{2}{3} = 0.66\bar{6}.$$

The percentage rounded to the nearest tenth place tells us that teachers work 66.7% of the year. Percentages also appear in real-world application problems involving finding missing quantities like in the following question: 60% of what number is 75? To find the missing quantity, an equation can be used. Let x be equal to the missing quantity. Therefore, $0.60x = 75$. Divide each side by 0.60 to obtain 125. Therefore, 60% of 125 is equal to 75.

Sales tax is an important application relating to percentages because tax rates are usually given as percentages. For example, a city might have an 8% sales tax rate. Therefore, when an item is purchased with that tax rate, the real cost to the customer is 1.08 times the price in the store. For example, a \$25 pair of jeans costs the customer $\$25 \times 1.08 = \$27$. Sales tax rates can also be determined if they are unknown when an item is purchased. If a customer visits a store and purchases an item for \$21.44, but the price in the store was \$19, they can find the tax rate by first subtracting $\$21.44 - \$19$ to obtain \$2.44, the sales tax amount. The sales tax is a percentage of the in-store price. Therefore, the tax rate is $\frac{2.44}{19} = 0.128$, which has been rounded to the nearest thousandths place. In this scenario, the actual sales tax rate given as a percentage is 12.8%.

Solving Real-World Problems Involving Proportions

Fractions appear in everyday situations, and in many scenarios, they appear in the real-world as ratios and in proportions. A **ratio** is formed when two different quantities are compared. For example, in a group of 50 people, if there are 33 females and 17 males, the ratio of females to males is 33 to 17. This expression can be written in the fraction form as $\frac{33}{50}$, where the denominator is the sum of females and males, or by using the ratio symbol, 33:17. The order of the number matters when forming ratios. In the same setting, the ratio of males to females is 17 to 33, which is equivalent to $\frac{17}{50}$ or 17:33. A **proportion** is an equation involving two ratios. The equation $\frac{a}{b} = \frac{c}{d}$, or $a{:}b = c{:}d$ is a proportion, for real numbers a, b, c, and d. Usually, in one ratio, one of the quantities is unknown, and cross-multiplication is used to solve for the unknown. Consider $\frac{1}{4} = \frac{x}{5}$. To solve for x, cross-multiply to obtain $5 = 4x$. Divide each side by 4 to obtain the solution $x = \frac{5}{4}$. It is also true that percentages are ratios in which the second term is 100 minus the first term. For example, 65% is 65:35 or $\frac{65}{100}$. Therefore, when working with percentages, one is also working with ratios.

Real-world problems frequently involve proportions. For example, consider the following problem: If 2 out of 50 pizzas are usually delivered late from a local Italian restaurant, how many would be late out of 235 orders? The following proportion would be solved with x as the unknown quantity of late pizzas: $\frac{2}{50} = \frac{x}{235}$. Cross multiplying results in $470 = 50x$. Divide both sides by 50 to obtain $x = \frac{470}{50}$, which in lowest terms is equal to $\frac{47}{5}$. In decimal form, this improper fraction is equal to 9.4. Because it does not make sense to answer this question with decimals (portions of pizzas do not get delivered) the answer must be rounded. Traditional rounding rules would say that 9 pizzas would be expected to be delivered late. However, to be safe, rounding up to 10 pizzas out of 235 would probably make more sense.

Solving Real-World Problems Involving Ratios and Rates of Change

Recall that a ratio is the comparison of two different quantities. Comparing 2 apples to 3 oranges results in the ratio 2:3, which can be expressed as the fraction $\frac{2}{5}$. Note that order is important when discussing ratios. The number mentioned first is the antecedent, and the number mentioned second is the consequent. Note that the consequent of the ratio and the denominator of the fraction are *not* the same. When there are 2 apples to 3 oranges, there are five fruit total; two fifths of the fruit are apples, while three fifths are oranges. The ratio 2:3 represents a different relationship that the ratio 3:2. Also, it is important to make sure that when discussing ratios that have units attached to them, the two quantities use the same units. For example, to think of 8 feet to 4 yards, it would make sense to convert 4 yards to feet by multiplying by 3. Therefore, the ratio would be 8 feet to 12 feet, which can be expressed as the fraction $\frac{8}{20}$. Also, note that it is proper to refer to ratios in lowest terms. Therefore, the ratio of 8 feet to 4 yards is equivalent to the fraction $\frac{2}{5}$.

Many real-world problems involve ratios. Often, problems with ratios involve proportions, as when two ratios are set equal to find the missing amount. However, some problems involve deciphering single ratios. For example, consider an amusement park that sold 345 tickets last Saturday. If 145 tickets were sold to adults and the rest of the tickets were sold to children, what would the ratio of the number of adult tickets to children's tickets be? A common mistake would be to say the ratio is 145:345. However, 345 is the total number of tickets sold, not the number of children's tickets. There were $345 - 145 = 200$ tickets sold to children. The correct ratio of adult to children's tickets is 145:200. As a fraction, this expression is written as $\frac{145}{345}$, which can be reduced to $\frac{29}{69}$.

While a ratio compares two measurements using the same units, rates compare two measurements with different units. Examples of rates would be $200 for 8 hours of work, or 500 miles traveled per 20 gallons. Because the units are different, it is important to always include the units when discussing rates. Rates can be easily seen because if they are expressed in words, the two quantities are usually split up using one of the following words: *for, per, on, from, in.* Just as with ratios, it is important to write rates in lowest terms. A common rate that can be found in many real-life situations is cost per unit. This quantity describes how much one item or one unit costs. This rate allows the best buy to be determined, given a couple of different sizes of an item with different costs. For example, if 2 quarts of soup was sold for $3.50 and 3 quarts was sold for \$4.60, to determine the best buy, the cost per quart should be found. $\frac{\$3.50}{2\text{ qt}} = \$1.75$ per quart, and $\frac{\$4.60}{3\text{ qt}} = \$1.53$ per quart. Therefore, the better deal would be the 3-quart option.

Rate of change problems involve calculating a quantity per some unit of measurement. Usually the unit of measurement is time. For example, meters per second is a common rate of change. To calculate this measurement, find the amount traveled in meters and divide by total time traveled. The calculation is an average of the speed over the entire time interval. Another common rate of change used in the real world is miles per hour. Consider the following problem that involves calculating an average rate of change in temperature. Last Saturday, the temperature at 1:00 a.m. was 34 degrees Fahrenheit, and at noon, the temperature had increased to 75 degrees Fahrenheit. What was the average rate of change over that time interval? The average rate of change is calculated by finding the change in temperature and dividing by the total hours elapsed. Therefore, the rate of change was equal to $\frac{75-34}{12-1} = \frac{41}{11}$ degrees per hour. This quantity rounded to two decimal places is equal to 3.72 degrees per hour.

A common rate of change that appears in algebra is the slope calculation. Given a linear equation in one variable, $y = mx + b$, the *slope*, m, is equal to $\frac{rise}{run}$ or $\frac{change\ in\ y}{change\ in\ x}$. In other words, slope is equivalent to the ratio of the vertical and horizontal changes between any two points on a line. The vertical change is known as the *rise*, and the horizontal change is known as the *run*. Given any two points on a line (x_1, y_1) and (x_2, y_2), slope can be calculated with the formula:

$$m = \frac{y_2 - y_1}{x_2 - x_1} = \frac{\Delta y}{\Delta x}$$

Common real-world applications of slope include determining how steep a staircase should be, calculating how steep a road is, and determining how to build a wheelchair ramp.

Many times, problems involving rates and ratios involve proportions. A proportion states that two ratios (or rates) are equal. The property of cross products can be used to determine if a proportion is true, meaning both ratios are equivalent. If $\frac{a}{b} = \frac{c}{d}$, then to clear the fractions, multiply both sides by the least common denominator, bd. This results in $ad = bc$, which is equal to the result of multiplying along both diagonals. For example, $\frac{4}{40} = \frac{1}{10}$ grants the cross product $4 \times 10 = 40 \times 1$, which is equivalent to $40 = 40$ and shows that this proportion is true. Cross products are used when proportions are involved in real-world problems. Consider the following: If 3 pounds of fertilizer will cover 75 square feet of grass, how many pounds are needed for 375 square feet? To solve this problem, a proportion can be set up using two ratios. Let x equal the unknown quantity, pounds needed for 375 feet. Then, the equation found by setting the two given ratios equal to one another is $\frac{3}{75} = \frac{x}{375}$. Cross-multiplication gives $3 \times 375 = 75x$. Therefore, $1{,}125 = 75x$. Divide both sides by 75 to get $x = 15$. Therefore, 15 pounds of fertilizer are needed to cover 375 square feet of grass.

Another application of proportions involves similar triangles. If two triangles have the same measurement as two triangles in another triangle, the triangles are said to be **similar.** If two are the same, the third pair of angles are equal as well because the sum of all angles in a triangle is equal to 180 degrees. Each pair of equivalent angles are known as **corresponding angles. Corresponding sides** face the corresponding angles, and it is true that corresponding sides are in proportion. For example, consider the following set of similar triangles:

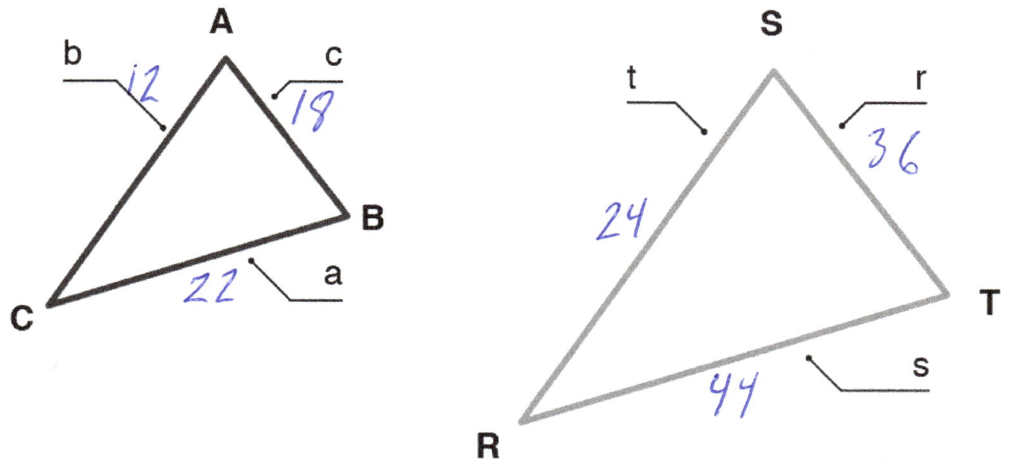

Angles A and S have the same measurement, angles C and R have the same measurement, and angles B and T have the same measurement. Therefore, the following proportion can be set up from the sides:

$$\frac{c}{r} = \frac{a}{s} = \frac{b}{t}$$

This proportion can be helpful in finding missing lengths in pairs of similar triangles. For example, if the following triangles are similar, a proportion can be used to find the missing side lengths, a and b.

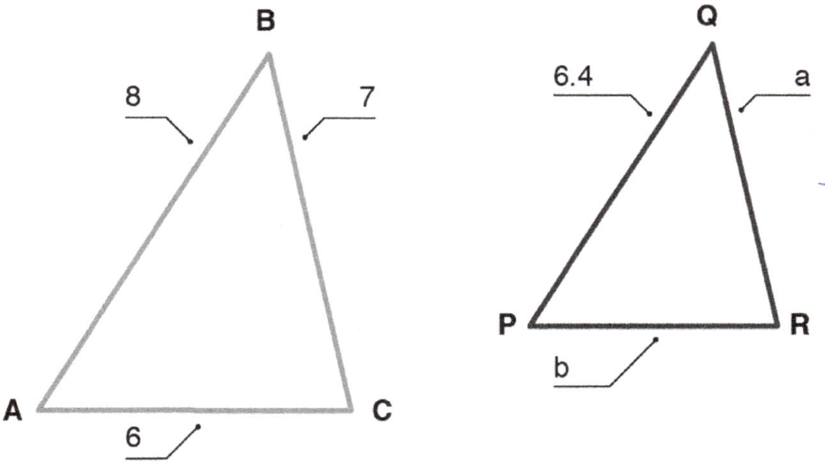

The proportions $\frac{8}{6.4} = \frac{6}{b}$ and $\frac{8}{6.4} = \frac{7}{a}$ can both be cross multiplied and solved to obtain $a = 5.6$ and $b = 4.8$.

A real-life situation that uses similar triangles involves measuring shadows to find heights of unknown objects. Consider the following problem: A building casts a shadow that is 120 feet long, and at the same time, another building that is 80 feet high casts a shadow that is 60 feet long. How tall is the first building? Each building, together with the sun rays and shadows casted on the ground, forms a triangle. They are similar because each building forms a right angle with the ground, and the sun rays form equivalent angles. Therefore, these two pairs of angles are both equal. Because all angles in a triangle add up to 180 degrees, the third angles are equal as well. Both shadows form corresponding sides of the triangle, the buildings form corresponding sides, and the sun rays form corresponding sides. Therefore, the triangles are similar, and the following proportion can be used to find the missing building length:

$$\frac{120}{x} = \frac{60}{80}$$

Cross-multiply to obtain the cross products, $9600 = 60x$. Then, divide both sides by 60 to obtain $x = 160$. This solution means that the other building is 160 feet high.

Applying Estimation Strategies and Rounding Rules to Real-World Problems

Sometimes it is helpful to find an estimated answer to a problem rather than working out an exact answer. An estimation might be much quicker to find, and given the scenario, an estimation might be all that is required. For example, if Aria goes grocery shopping and has only a $100 bill to cover all of her purchases, it might be appropriate for her to estimate the total of the items she is purchasing to determine if she has enough money to cover them. Also, an estimation can help determine if an answer makes sense. For example, if an answer in the 100s is expected, but the result is a fraction less than 1, something is wrong in the calculation.

The first type of estimation involves rounding. **Rounding** consists of expressing a number in terms of the nearest decimal place like the tenth, hundredth, or thousandth place, or in terms of the nearest whole number unit like tens, hundreds, or thousands place. When rounding to a specific place value, look at the digit to the right of the place. If it is 5 or higher, round the number to its left up to the next value, and if it is 4 or lower, keep that number at the same value. For instance, 1,654.2674 rounded to the nearest thousand is 2,000, and the same number rounded to the nearest thousandth is 1,654.267. Rounding can be used in the scenario when grocery totals need to be estimated. Items can be rounded to the nearest dollar. For example, a can of corn that costs $0.79 can be rounded to $1.00, and then all other items can be rounded in a similar manner and added together. When working with larger numbers, it might make more sense to round to higher place values. For example, when estimating the total value of a dealership's car inventory, it would make sense to round the car values to the nearest thousands place. The price of a car that is on sale for $15,654 can be estimated at $16,000.

All other cars on the lot could be rounded in the same manner, and then their sum can be found. Depending on the situation, it might make sense to calculate an over-estimate. For example, to make sure Aria has enough money at the grocery store, rounding up every time for each item would ensure that she will have enough money when it comes time to pay. A $0.40 item rounded up to $1.00 would ensure that there is a dollar to cover that item. Traditional rounding rules would round $0.40 to $0, which does not make sense in this particular real-world setting. Aria might not have a dollar available at checkout to pay for that item if she uses traditional rounding. It is up to the customer to decide the best approach when estimating.

Estimating is also very helpful when working with measurements. Bryan is updating his kitchen and wants to retile the floor. Again, an over-measurement might be useful. Also, rounding to nearest half-unit might be helpful. For instance, one side of the kitchen might have an exact measurement of 14.32 feet, and the most useful measurement needed to buy tile could be estimating this quantity to be 14.5 feet. If the kitchen was rectangular and the other side measured 10.9 feet, Bryan might round the other side to 11 feet.

Therefore, Bryan would find the total tile necessary according to the following area calculation:

$$14.5 \times 11 = 159.5 \text{ square feet}$$

To make sure he purchases enough tile, Bryan would probably want to purchase at least 160 square feet of tile. This is a scenario in which an estimation might be more useful than an exact calculation. Having more tile than necessary is better than having an exact amount, in case any tiles are broken or otherwise unusable.

Finally, estimation is helpful when exact answers are necessary. Consider a situation in which Sabina has many operations to perform on numbers with decimals, and she is allowed a calculator to find the result. Even though an exact result can be obtained with a calculator, there is always a possibility that Sabina could make an error while inputting the data. For example, she could miss a decimal place, or misuse a parenthesis, causing a problem with the actual order of operations. In this case, a quick estimation at the beginning would be helpful to make sure the final answer is given with the correct number of units. Sabina has to find the exact total of 10 cars listed for sale at the dealership. Each price has two decimal places included to account for both dollars and cents. If one car is listed at $21,234.43 but Sabina incorrectly inputs into the calculator the price of $2,123.443, this error would throw off the final sum by almost $20,000. A quick estimation at the beginning, by rounding each price to the nearest thousands place and finding the sum of the prices, would give Sabina an amount to compare the exact amount to. This comparison would let Sabina see if an error was made in her exact calculation.

Algebraic Reasoning

Quadratic and Other Polynomial Expressions, Equations, and Functions

In math, a **relation** is a relationship between two sets of numbers. By using a rule, it takes a number from the first set and matches it to a number in the second set. A relation consists of a set of inputs, known as the **domain,** and a set of outputs, known as the **range.** A **function** is a relation in which each member of the domain is paired to only one other member of the range. In other words, each input has only one output.

Here is an example of a relation that is not a function:

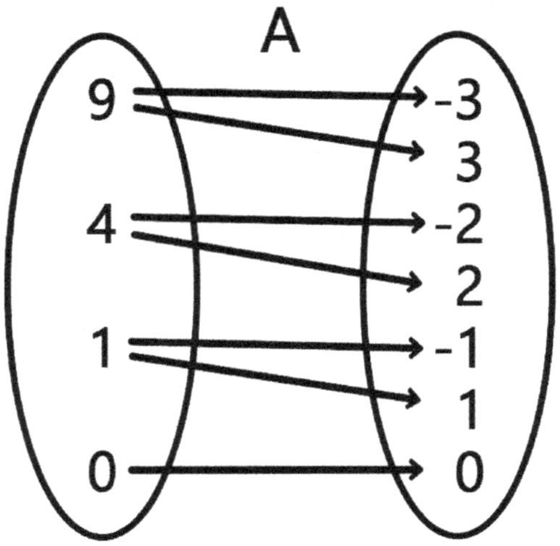

Every member of the first set, the domain, is mapped to two members of the second set, the range. Therefore, this relation is not a function.

A function can also be represented by a table of ordered pairs, a graph of ordered pairs (a scatterplot), or a set of ordered pairs as shown in the following:

Mapping

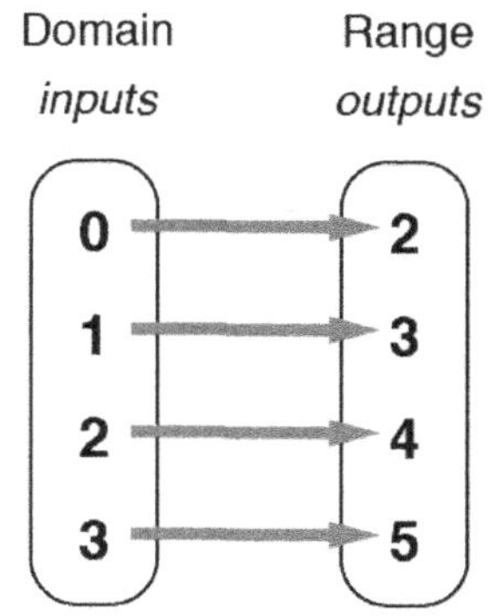

Table

x	y
0	2
1	3
2	4
3	5

Graph

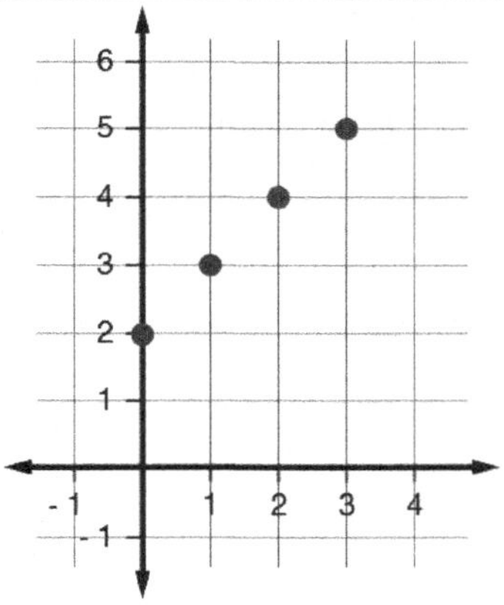

Ordered Pairs

{(0,2),(1,3),(2,4),(3,5)}

Note that this relation is a function because every member of the domain is mapped to exactly one member of the range.

An **equation** occurs when two algebraic expressions are set equal to one another. Functions can be represented in equation form. Given an equation in two variables, *x* and *y*, it can be expressed in function form if solved for *y*. For example, the linear equation $2x + y = 5$ can be solved for *y* to obtain $y = -2x + 5$, otherwise known as **slope-intercept** form. To place the equation in function form, replace *y* with *f(x)*, which is read "*f* of *x*." Therefore:

$$f(x) = -2x + 5$$

This notation clarifies the input–output relationship of the function. The function *f* is a function of *x*, so an *x* value can be plugged into the function to obtain an output. For example:

$$f(2) = -2 \times 2 + 5 = 1$$

Therefore, an input of 2 corresponds to an output of 1.

A function can be graphed by plotting ordered pairs in the *xy*-plane in the same way that the equation form is graphed. The graph of a function always passes the **Vertical Line Test.** Basically, for any graph, if a vertical line can be drawn through any part of the graph and it hits the graph in more than one place, the graph is not a function. For example, the graph of a circle is not a function. The Vertical Line Test shows that with these relationships, the same *x* value has more than one *y* value, which goes against the definition of a function.

Inequalities look like equations, but instead of an equals sign, $<, >, \leq, \geq$, or $\neq$ are used. Here are some examples of inequalities: $2x + 7 < y, 3x^2 \geq 5$, and $x \neq 4$. Inequalities show relationships between algebraic expressions when the quantities are different. Inequalities can also be expressed in function form if they are solved for *y*. For instance, the first inequality listed above can be written as:

$$2x + y < f(x).$$

Even and Odd Functions

A function is considered **even** when $f(x) = f(-x)$ for all values of x. This relationship means that the graph of an even function is perfectly symmetrical about the y-axis. In other words, the graph is reflected over the y-axis. The term "even" describes these functions because functions like x^2, x^4, x^6, and so on display this characteristic. That said, there are some functions that are even that don't involve x raised to an even exponent. For example, the graph of $f(x) = cos(x)$ is also symmetrical over the y-axis. There are also function that *do* have an even exponent that are *not* even functions; although $f(x) = x^2 + 3$ is an even function, $f(x) = (x + 3)^2$ is not because $f(x) \neq f(-x)$ for all values of x.

A function is considered **odd** when -$f(x) = f(-x)$ for all values of x. This relationship means that the graph of an odd function is perfectly symmetrical about the origin. The term "odd" describes these functions because functions like x, x^3, x^5, x^7, and so on display this characteristic. As with even functions, exceptions to these exponential values exist. For example, the graph of $f(x) = sin(x)$ also has origin symmetry. There are also function that *do* have an odd exponent that are *not* odd functions; while $f(x) = x^3 - 3x$ is an odd function, $f(x) = x^3 - 1$ is not because -$f(x) \neq f(-x)$ for all values of x.

The majority of functions do not display these unique relationships and are thus neither even nor odd. Essentially, to algebraically determine if a function is even, odd, or neither, various values need to be plugged in for $f(x)$ and $f(-x)$. If all values of x yield the same output for $f(x)$ and $f(-x)$, the function is even. If all values of x yield the same output for -$f(x)$ and $f(-x)$, the function is odd. Any other situation indicates the function is neither even nor odd. To graphically determine if a function is even, odd, or neither, symmetry needs to be evaluated. If the function is perfectly symmetrical across the y-axis, it is an even function. If it is symmetrical about the origin, the function is odd. Any other situation indicates that the function is neither even nor odd.

It should be noted that there is one function that is both even *and* odd: $f(x) = 0$.

Quadratic Functions

When given data in ordered pairs, choosing an appropriate function or equation to model the data is important. Besides linear relationships, other common relationships that exist are quadratic and exponential. A helpful way to determine what type of function to use is to find the difference between consecutive dependent variables. Basically, find pairs of ordered pairs where the *x*-values increase by 1, and take the difference of the *y*-values. If the differences in the *y*-values are always the same value, then the function is **linear**. If the differences in the *y*-values when the *x*-values increase by 1 are not the same, the function could be quadratic or exponential. If the differences are not the same, find differences of those differences. If consecutive differences are the same, then the function is **quadratic**. If consecutive differences are not the same, try taking ratios of consecutive *y*-values. If the ratios are the same, the data have an exponential relationship and an exponential function should be used.

For example, the ordered pairs (1, 4), (2, 6), (3 ,8), and (4,10) have a linear relationship because the difference in *y*-values is 2 for every increase in *x* of 1. The ordered pairs (1, 0), (2, 3), (3, 10), and (4, 21) have a nonlinear relationship. The first differences in *y*-values are 3, 7, and 11; however, consecutive second differences are both 4, which means this is a quadratic relationship. Lastly, the ordered pairs (1, 10), (2, 30), (3, 90), and (4,270) have an exponential relationship. Taking ratios of consecutive *y*-values leads to a common ratio of 4.

The general form of a **quadratic equation** is $y = ax^2 + bx + c$, and its vertex form is $y = a(x - h)^2 + k$, with vertex (h, k). If the vertex and one other point are known, the vertex form should be used to solve for *a*. If three points, not the vertex, are known, the general form should be used. The three points create a system of three equations in three unknowns that can be solved for.

Solving a Quadratic Equation

Given a quadratic equation in standard form, $ax^2 + bx + c = 0$, with constants *a*, *b*, and *c*, such that $c \neq 0$, it can have either two real solutions, one real solution, or two complex solutions of the form $a + bi$ (no real solutions). The number of solutions is determined using its determinant $b^2 - 4ac$. If the determinant is positive, there are two real solutions. If the determinant is negative, there are no real solutions. If the determinant is equal to 0, there is one real solution. For example, given the quadratic equation $2x^2 - x + 4 = 0$, its determinant is:

$$(-1)^2 - 4(2)(4) = 1 - 32 = -31$$

which is less than 0. Therefore, it has two complex solutions.

There are a number of ways to solve a quadratic equation. The first way is through **factoring.** If the equation is in standard form and the polynomial can be factored, set each factor equal to 0 and solve. This can be done because if $ab = 0$, either $a = 0, b = 0$, or both are equal to 0. For example:

$$x^2 - 7x + 10 = (x - 5)(x - 2)$$

Therefore, the solutions of $x^2 - 7x + 10 = 0$ are those that satisfy both $x - 5 = 0$ and $x - 2 = 0$, or $x = 5$ and 2. This is the simplest method to solve quadratic equations; however, not all quadratic polynomials can be factored, so this method does not work for all quadratic equations.

Another method is through **completing the square**. The polynomial $x^2 + 10x - 9$ cannot be factored, so complete the square in the equation $x^2 + 10x - 9 = 0$ to find its solutions. First, add 9 to both sides, resulting in:

$$x^2 + 10x = 9$$

Then, divide the *x*-coefficient by 2, square it, and add it to both sides of the equation. In this example, $\left(\frac{10}{2}\right)^2 = 25$ is added to both sides of the equation to obtain:

$$x^2 + 10x + 25 = 9 + 25 = 34$$

The polynomial, which is now a perfect square trinomial, can then be factored into $(x+5)^2 = 34$. Finally, solving for x involves first taking the square root of both sides and then subtracting 5 from both sides. This process leads to the two solutions:

$$x = \pm\sqrt{34} - 5$$

This method always works for any quadratic equation.

The final method of solving a quadratic equation is to use the **quadratic formula**. Given a quadratic equation in standard form, $ax^2 + bx + c = 0$, its solutions always can be found using the formula:

$$x = \frac{-b \pm \sqrt{b^2 - 4ac}}{2a}$$

This method, like completing the square, can always be used.

Polynomial Functions

A **polynomial function** is a function containing a polynomial expression, which is an expression containing constants and variables combined using the four mathematical operations. The degree of a polynomial depends on the largest exponent in the expression. Typical polynomial functions are **quartic,** with a degree of 4, **cubic,** with a degree of 3, and **quadratic,** with a degree of 2. Note that the exponents on the variables can only be nonnegative integers. The domain of any polynomial function is all real numbers because any number plugged into a polynomial expression grants a real number output. An example of a quartic polynomial equation is:

$$y = x^4 + 3x^3 - 2x + 1$$

The zeros of a polynomial function are the points where its graph crosses the y-axis. In order to find the number of real zeros of a polynomial function, use **Descartes' Rule of Signs**, which states that the number of possible positive real zeros is equal to the number of sign changes in the coefficients. If there is only one sign change, there is only one positive real zero. In the example above, the signs of the coefficients are positive, positive, negative, and positive. The sign changes twice; therefore, there are at most two positive real zeros. The number of possible negative real zeros is equal to the number of sign changes in the coefficients when plugging $-x$ into the equation. Again, if there is only one sign change, there is only one negative real zero. The polynomial result when plugging -x into the equation is:

$$y = (-x)^4 + 3(-x)^3 - 2(-x) + 1$$

$$y = x^4 - 3x^3 + 2x + 1$$

The sign changes two times, so there are at most two negative real zeros. Another polynomial equation this rule can be applied to is:

$$y = x^3 + 2x - x - 5$$

There is only one sign change in the terms of the polynomial, so there is exactly one real zero. When plugging -x into the equation, the polynomial result is:

$$-x^3 - 2x - x - 5$$

There are no sign changes in this polynomial, so there are no possible negative zeros.

Adding, Subtracting, and Multiplying Polynomial Equations

When working with polynomials, **like terms** are terms that contain exactly the same variables with the same powers. For example, x^4y^5 and $9x^4y^5$ are like terms. The coefficients are different, but the same variables are raised to the same powers. When adding polynomials, only terms that are like can be added. When adding two like terms, just add the coefficients and leave the variables alone. This process uses the distributive property. For example:

$$x^4y^5 + 9x^4y^5$$

$$(1 + 9)x^4y^5$$

$$10x^4y^5$$

Therefore, when adding two polynomials, simply add the like terms together. Unlike terms cannot be combined.

Subtracting polynomials involves adding the opposite of the polynomial being subtracted. Basically, the sign of each term in the polynomial being subtracted is changed, and then the like terms are combined because it is now an addition problem. For example, consider the following:

$$6x^2 - 4x + 2 - (4x^2 - 8x + 1).$$

Add the opposite of the second polynomial to obtain:

$$6x^2 - 4x + 2 + (-4x^2 + 8x - 1)$$

Then, collect like terms to obtain:

$$2x^2 + 4x + 1.$$

Multiplying polynomials involves using the product rule for exponents that:

$$b^m b^n = b^{m+n}$$

Basically, when multiplying expressions with the same base, just add the exponents. Multiplying a monomial times a monomial involves multiplying the coefficients together and then multiplying the variables together using the product rule for exponents. For instance:

$$8x^2y \times 4x^4y^2 = 32x^6y^3$$

When multiplying a monomial times a polynomial that is not a monomial, use the distributive property to multiply each term of the polynomial times the monomial.

For example:

$$3x(x^2 + 3x - 4)$$

$$3x^3 + 9x^2 - 12x$$

Finally, multiplying two polynomials when neither one is a monomial involves multiplying each term of the first polynomial times each term of the second polynomial. There are some shortcuts, given certain scenarios.

For instance, a binomial times a binomial can be found by using the **FOIL (Firsts, Outers, Inners, Lasts)** method shown here:

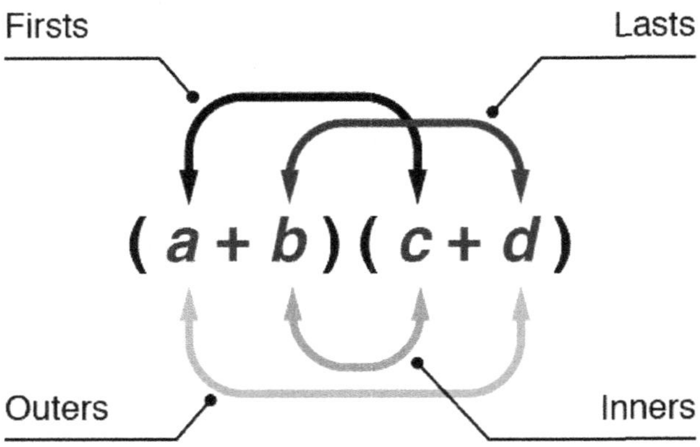

Finding the product of a sum and difference of the same two terms is simple because if it was to be foiled out, the outer and inner terms would cancel out. For instance:

$$(x + y)(x - y) = x^2 + xy - xy - y^2$$

Finally, the square of a binomial can be found using the following formula:

$$(a \pm b)^2 = a^2 \pm 2ab + b^2$$

The Relationship Between Zeros and Factors of Polynomials

A polynomial is a mathematical expression containing addition, subtraction, or multiplication of one or more constants multiplied by variables raised to positive powers. A **polynomial equation** is a polynomial set equal to another polynomial, or in standard form, a polynomial is set equal to zero. A **polynomial function** is a polynomial set equal to y. For instance, $x^2 + 2x - 8$ is a polynomial, $x^2 + 2x - 8 = 0$ is a polynomial equation, and $y = x^2 + 2x - 8$ is the corresponding polynomial function. To solve a polynomial equation, the x-values in which the graph of the corresponding polynomial function crosses the x-axis are sought.

These coordinates are known as the **zeros** of the polynomial function because they are the coordinates in which the y-coordinates are 0. One way to find the zeros of a polynomial is to find its factors, then set each individual factor equal to 0, and solve each equation to find the zeros. A **factor** is a linear expression, and to completely factor a polynomial, the polynomial must be rewritten as a product of individual linear factors. The polynomial listed above can be factored as $(x + 4)(x - 2)$. Setting each factor equal to zero results in the zeros $x = -4$ and $x = 2$.

Here is the graph of the zeros of the polynomial:

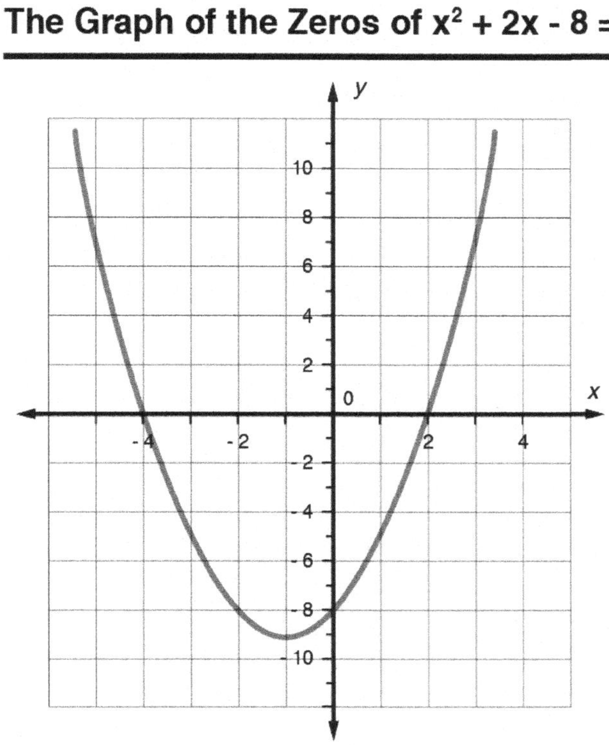

Expressions, Equations and Functions Involving Powers, Roots, and Radicals

Numbers can also be written using exponents. The number 7,000 can be written as $7 \times 1{,}000$ because 7 is in the thousands place. It can also be written as 7×10^3 because $1{,}000 = 10^3$. Another number that can use this notation is 500. It can be written as 5×100, or 5×10^2, because $100 = 10^2$. The number 30 can be written as 3×10, or 3×10^1, because $10 = 10^1$. Notice that each one of the exponents of 10 is equal to the number of zeros in the number. Seven is in the thousands place, with three zeros, and the exponent on ten is 3. The five is in the hundreds place, with two zeros, and the exponent on the ten is 2. A question may give the number 40,000 and ask for it to be rewritten using exponents with a base of ten. Because the number has a four in the ten-thousands place and four zeros, it can be written using an exponent of four: 4×10^4.

The *n*th root of *a* is given as $\sqrt[n]{a}$, which is called a **radical.** Typical values for *n* are 2 and 3, which represent the square and cube roots. In this form, *n* represents an integer greater than or equal to 2, and *a* is a real number. If *n* is even, *a* must be nonnegative, and if *n* is odd, *a* can be any real number. This radical can be written in exponential form as $a^{\frac{1}{n}}$. Therefore, $\sqrt[4]{15}$ is the same as $15^{\frac{1}{4}}$ and $\sqrt[3]{-5}$ is the same as $(-5)^{\frac{1}{3}}$.

In a similar fashion, the *n*th root of *a* can be raised to a power *m*, which is written as $\left(\sqrt[n]{a}\right)^m$. This expression is the same as $\sqrt[n]{a^m}$. For example:

$$\sqrt[2]{4^3} = \sqrt[2]{64} = 8 = \left(\sqrt[2]{4}\right)^3 = 2^3$$

Because $\sqrt[n]{a} = a^{\frac{1}{n}}$, both sides can be raised to an exponent of *m*, resulting in:

$$\left(\sqrt[n]{a}\right)^m = \sqrt[n]{a^m} = a^{\frac{m}{n}}$$

This rule allows:

$$\sqrt[2]{4^3} = \left(\sqrt[2]{4}\right)^3 = 4^{\frac{3}{2}} = (2^2)^{\frac{3}{2}} = 2^{\frac{6}{2}} = 2^3 = 8$$

Negative exponents can also be incorporated into these rules. Any time an exponent is negative, the base expression must be flipped to the other side of the fraction bar and rewritten with a positive exponent. For example, $2^{-3} = \frac{1}{2^3} = \frac{1}{8}$. Therefore, two more relationships between radical and exponential expressions are:

$$a^{-\frac{1}{n}} = \frac{1}{\sqrt[n]{a}}$$

$$a^{-\frac{m}{n}} = \frac{1}{\sqrt[n]{a^m}} = \frac{1}{\left(\sqrt[n]{a}\right)^m}$$

Thus:

$$8^{-\frac{1}{3}} = \frac{1}{\sqrt[3]{8}} = \frac{1}{2}$$

All of these relationships are very useful when simplifying complicated radical and exponential expressions. If an expression contains both forms, use one of these rules to change the expression to contain either all radicals or all exponential expressions. This process makes the entire expression much easier to work with, especially if the expressions are contained within equations.

Consider the following example: $\sqrt{x} \times \sqrt[4]{x}$. It is written in radical form; however, it can be simplified into one radical by using exponential expressions first. The expression can be written as $x^{\frac{1}{2}} \times x^{\frac{1}{4}}$. It can be combined into one base by adding the exponents as:

$$x^{\frac{1}{2}+\frac{1}{4}} = x^{\frac{3}{4}}$$

Writing this back in radical form, the result is $\sqrt[4]{x^3}$.

Creating Equivalent Expressions Involving Rational Exponents and Radicals

Re-writing complex radical expressions as equivalent forms with rational exponents can help to simplify them. The rule that helps this conversion is:

$$\sqrt[n]{x^m} = x^{\frac{m}{n}}$$

If $m = 1$, the rule is simply $\sqrt[n]{x} = x^{\frac{1}{n}}$. For example, consider the following expression: $\sqrt[4]{x}\sqrt[2]{y}$. This can be written as one radical expression, but first it needs to be converted to an equivalent expression. The equivalent expression is $x^{\frac{1}{4}}y^{\frac{1}{2}}$. The goal is to have one radical, which means one index *n*, so a common denominator of the exponents must be found. The common denominator is 4, so an equivalent expression is $x^{\frac{1}{4}}y^{\frac{2}{4}}$. The exponential rule $a^m b^m = (ab)^m$ can be used to factor $\frac{1}{4}$ out of both variables. This process results in the expression $(xy^2)^{\frac{1}{4}}$, and its equivalent radical form is $\sqrt[4]{xy^2}$. Converting to rational exponents has allowed the entire expression to be written as one radical.

Another type of problem could involve going in the opposite direction: starting with rational exponents and using an equivalent radical form to simplify the expression. For example, $32^{\frac{1}{5}}$ might not appear equal to 2, but putting it in its equivalent radical form $\sqrt[5]{32}$ shows that it is equivalent to the fifth root of 32, which is 2.

Radical Functions

A **radical equation** is an equation that contains a variable in the **radicand,** which is the expression under the root. The radical can be a square root, a cube root, or a higher root. To solve an equation containing a root, arrange the terms so that one radical is by itself on one side of the equals sign. Then raise both sides of the equation to the value of the root. For example, square both sides if the root contains a square root, and cube both sides if the root contains a cube root. Then, solve the resulting equation. If the equation still contains a radical, those steps must be completed again to remove the equation of all radicals. Finally, it is crucial that all solutions are checked in the original equation. Some solutions to this equation, once the radicals are removed, might not be solutions to the original radical equation. Consider the following radical equation:

$$\sqrt{3x+1} - \sqrt{x+4} = 1$$

Add the second radical to both sides, and then square both sides to obtain:

$$3x + 1 = x + 4 + 2\sqrt{x+4} + 1.$$

Next, collect like terms and isolate the radical to obtain:

$$4x^2 - 16x + 16 = 4(x + 4).$$

This simplifies into the quadratic equation $4x^2 - 20x$, which can be solved using factoring:

$$4x(x - 5) = 0.$$

So, it has solutions, $x = 0$ and $x = 5$. Both values must be checked into the original radical equation. because $x = 0$ does not check, it is not a real solution and is called an **extraneous solution**. However, $x = 5$ is a solution.

If an equation contains a variable in the denominator of a fraction, it is known as a **rational equation.** Anything, that when plugged into the equation, contains a zero denominator, cannot be a solution of the equation. To solve a rational equation, multiply both sides of the equation by the LCD (least common denominator) of all of the terms in the equation. Then, solve the resulting equation, making sure that the solutions do not cause any term to have a zero denominator in the original equation. Here is an example of solving the rational equation:

$$\frac{5}{x} - \frac{1}{3} = \frac{1}{x}$$

$$3x \times \left(\frac{5}{x} - \frac{1}{3}\right) = 3x \times \left(\frac{1}{x}\right)$$

Multiply both sides by the LCD

$$3x \times \frac{5}{x} - 3x \times \frac{1}{3} = 3x \times \frac{1}{x}$$

Distribute

$$15 - x = 3$$

$$-x = -12$$

$$x = 12$$

Simplify, and then solve

Both sides of the equation were initially multiplied times the LCD, $3x$. Note that the only number that could cause a problem as a solution would be $x = 0$, because it would create a 0 in the denominator.

Rational and Exponential Expressions, Equations, and Functions

Rational Expressions

A **rational expression** is a fraction or a ratio in which both the numerator and denominator are polynomials that are not equal to zero. A polynomial is a mathematical expression containing addition, subtraction, or multiplication of one or more constants multiplied by variables raised to positive powers. Here are some examples of rational expressions: $\frac{2x^2+6x}{x}$, $\frac{x-2}{x^2-6x+8}$, and $\frac{x^3-1}{x+2}$. Such expressions can be simplified using different forms of division. The first example can be simplified in two ways. Then, cancelling out an x in each numerator and the x in each denominator results in $2x + 6$. It also can be simplified using factoring and then crossing common factors out of the numerator and denominator. For example, it can be written as

$$\frac{2x(x+3)}{x} = 2(x+3) = 2x + 6$$

The second expression above can also be simplified using factoring. It can be written as:

$$\frac{x-2}{(x-2)(x-4)} = \frac{1}{x-4}$$

Finally, the third example can only be simplified using long division, as there are no common factors in the numerator and denominator. First, divide the first term of the numerator by the first term of the denominator, then write the result in the quotient. Then, multiply the divisor times that number and write it below the dividend. Subtract and continue the process until each term in the divisor is accounted for.

Here is the actual long division:

Simplifying Expressions Using Long Division

	x^2	$-2x$	$+4$	
$x+2$	x^3			-1
	x^3	$+2x^2$		
		$-2x^2$		-1
		$-2x^2$	$-4x$	
			$4x$	-1
			$4x$	$+8$
				-9

Exponential Equations

An **exponential equation** is used to model something with exponential growth or decay. If something grows exponentially, such as compound interest, the amount is multiplied times a growth factor for every increase in *x*. If something decays exponentially, the amount is multiplied times a factor between 0 and 1 for every increase in *x*. When a population is declining, an exponential decay equation can be used to represent the situation. The general form of an exponential function is $y = b \times a^x$, where *a* is the base and *b* is the y-intercept.

Exponential growth involves the dependent variable changing by a common ratio every unit increase. The equation of exponential growth is $y = a^x$ for $a > 0$, $a \neq 1$. The value a is known as the **base**. Consider the exponential equation $y = 2^x$. When x equals 1, y equals 2, and when x equals 2, y equals 4. For every unit increase in x, the value of the output variable doubles. Here is the graph of $y = 2^x$. Notice that as the dependent variable, y, gets very large, x increases slightly.

This characteristic of this graph is why sometimes a quantity is said to be blowing up exponentially.

$y = 2^x$

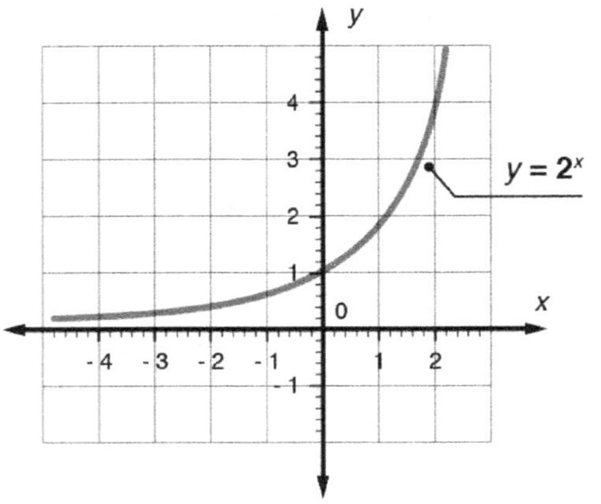

Logarithmic Functions

For $x > 0, b > 0, b \neq 1$, the function $f(x) = \log_b x$ is known as the **logarithmic function** with base b. With $y = \log_b x$, its exponential equivalent is $b^y = x$. In either case, the **exponent** is y and the **base** is b. Therefore, $3 = \log_2 8$ is the same as $2^3 = 8$. So, in order to find the logarithm with base 2 of 8, find the exponent that when 2 is raised to that value results in 8. Similarly, $\log_3 243 = 5$. In order to do this mentally, ask the question, what exponent does 3 need to be raised to that results in 243? The answer is 5. Most logarithms do not have whole number results. In this case, a calculator can be used. A calculator typically has buttons with base 10 and base e (Euler's number, the base in the natural log), so the change of base formula can be used to calculate these logs. For instance:

$$\log_3 55 = \frac{\log 55}{\log 3} = 3.64$$

Similarly, the natural logarithm with base e could be used to obtain the same result:

$$\log_3 55 = \frac{\ln 55}{\ln 3} = 3.64$$

The domain of a logarithmic function $f(x) = \log_b x$ is all positive real numbers. This is because the exponent must be a positive number. The range of a logarithmic function $f(x) = \log_b x$ is all real numbers. The graphs of all logarithmic functions of the form $f(x) = \log_b x$ always pass through the point (1, 0) because anything raised to the power of 0 is 1. Therefore, such a function always has an *x*-intercept at 1. If the base is greater than 1, the graph increases from the left to the right along the *x*-axis. If the base is between 0 and 1, the graph decreases from the left to the right along the *x*-axis. In both situations, the *y*-axis is a vertical asymptote.

The graph will never touch the *y*-axis, but it does approach it closely. Here are the graphs of the two cases of logarithmic functions:

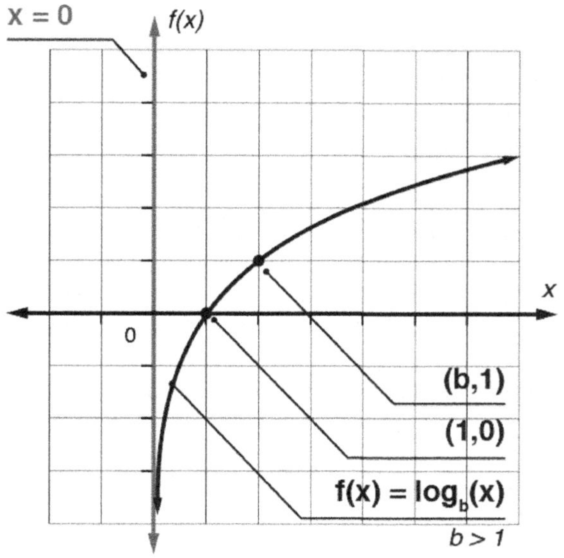

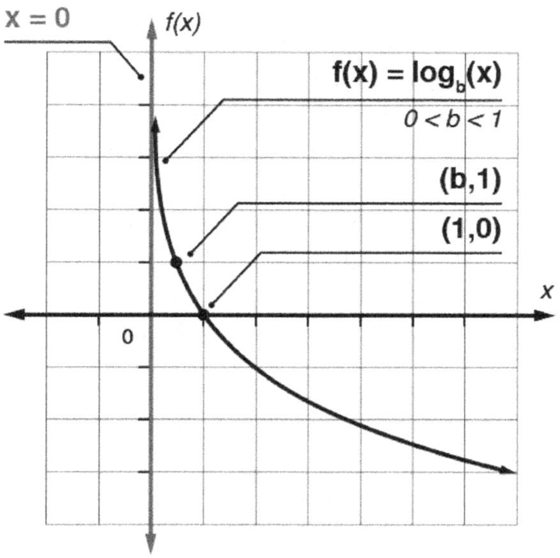

Geometric and Spatial Reasoning

Plane Geometry

Coordinate geometry is the intersection of algebra and geometry. Within this system, the points in a geometric shape are defined using ordered pairs. In the two-dimensional coordinate system, an x- and y-axis form the *xy-plane*. The x-axis is a horizontal scale, and the y-axis is a vertical scale. The ordered pair where the axes cross is known as the **origin**. To the right of the origin, the x values are positive, and to the left of the origin, the x values are negative. The y values above the origin are positive, and y values below the origin are negative. The axes split the plane into four quadrants, and the first quadrant is where both x and y values are positive. To plot an ordered pair means to locate the point corresponding to the x and y coordinates. For example, plotting (4,3) means moving to the right 4 units from 0 in the x direction and then moving up 3 units in the y direction.

Here is a picture of the xy-plane, also known as the **rectangular** or **Cartesian coordinate system**:

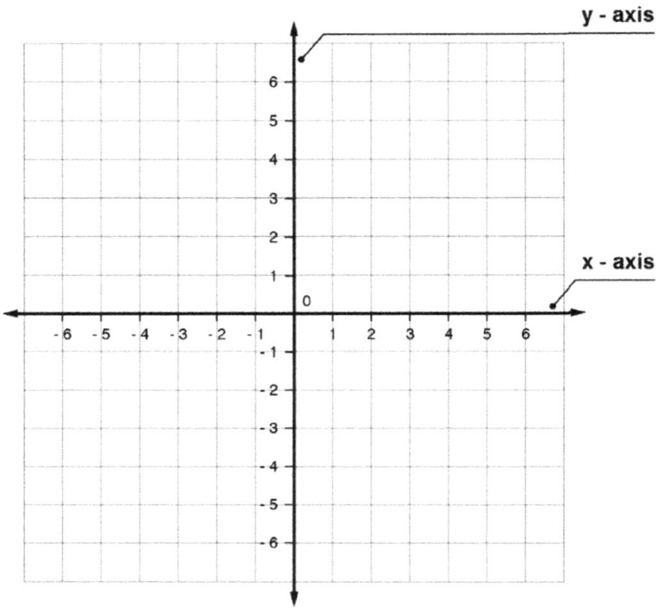

The coordinate system allows us to visualize relationships between equations and geometric figures. For instance, an equation in two variables, x and y, is represented as a straight line on the xy-coordinate plane. A solution of an equation in two variables is an ordered pair that satisfies the equation. A graph of an equation can be found by plotting several ordered pairs that are solutions of the equation and then connecting those points with a straight line or smooth curve. Here is the graph of:

$$4x + y = 8$$

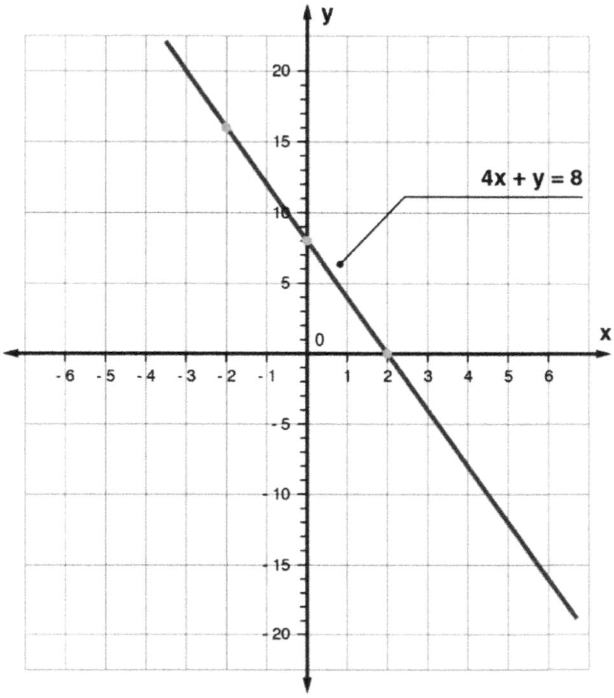

Three ordered pairs that are solutions to the equation were found and plotted. Those points are (–2,16), (0,8), and (2,0). The points were connected using a straight line. Note that the point (0,8) is where the line crosses the *y*-axis. This point is known as the ***y*-intercept**. The *y*-intercept can always be found by plugging $x = 0$ into the equation. Also, the point (2,0) is where the line crosses the *x*-axis. This point is known as the ***x*-intercept**, and it can always be found for any equation of a line by plugging $y = 0$ into the equation. The equation above is written in standard form, $Ax + By = C$. Often an equation is written in slope-intercept form, $y = mx + b$, where m represents the slope of the line, and b represents the *y*-intercept. The above equation can be solved for *y* to obtain $y = -4x + 8$, which shows a slope of –4 and a *y*-intercept of 8, meaning the point (0,8).

The **slope** of a line is the measure of steepness of a line, and it compares the vertical change of the line, the **rise**, to the horizontal change of the line, the **run**. The formula for slope of a line through two distinct points (x_1, y_1) and (x_2, y_2) is:

$$m = \frac{y_2 - y_1}{x_2 - x_1}$$

If the line increases from left to right, the slope is positive, and if the line decreases from left to right, as shown above, the slope is negative. If a line is horizontal, like the line representing the equation $y = 5$, the slope is 0. If a line is vertical, like the line representing the equation $x = 2$, the line has undefined slope.

In order to graph a function, it can be done the same way as equations. The $f(x)$ represents the dependent variable y in the equation, so replace $f(x)$ with y and plot some points. For example, the same graph above would be found for the function:

$$f(x) = -4x + 8$$

Graphs other than straight lines also exist. For example, here are the graphs of $f(x) = x^2$ and $f(x) = x^3$, the squaring and cubic functions.

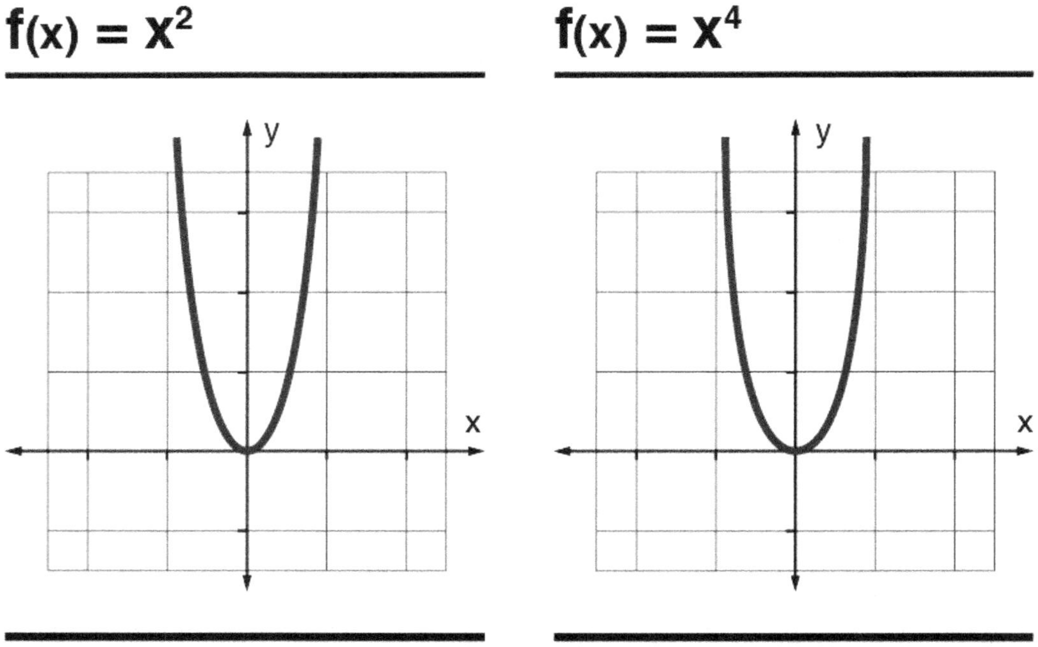

If the equals sign is changed to an inequality symbol such as $<. >, \leq$, or $\geq$ in an equation, the result is an inequality. If it is changed to a linear equation in two variables, the result is a linear inequality in two variables. A solution of an inequality in two variables is an ordered pair that satisfies the inequality. For example, (1,3) is a solution of the linear inequality $y \geq x + 1$ because when plugged in, it results in a true statement. The graph of an inequality in two variables consists of all ordered pairs that make the solution true. A half-plane consists of the set of all points on one side of a line in the *xy*-plane, and the solution to a linear inequality is a half-plane.

If the inequality consists of $>$ or $<$, the line is dashed, and no solutions actually exist on the line shown. If the inequality consists of $\geq$ or $\leq$, the line is solid, and solutions do exist on the line shown. In order to graph a linear inequality, graph the corresponding equation found by replacing the inequality symbol with an equals sign. Then pick a test point on either side of the line. If that point results in a true statement when plugged into the original inequality, shade in the side containing the test point. If it results in a false statement, shade in the opposite side.

Here is the graph of the inequality:

$$y < x + 1.$$

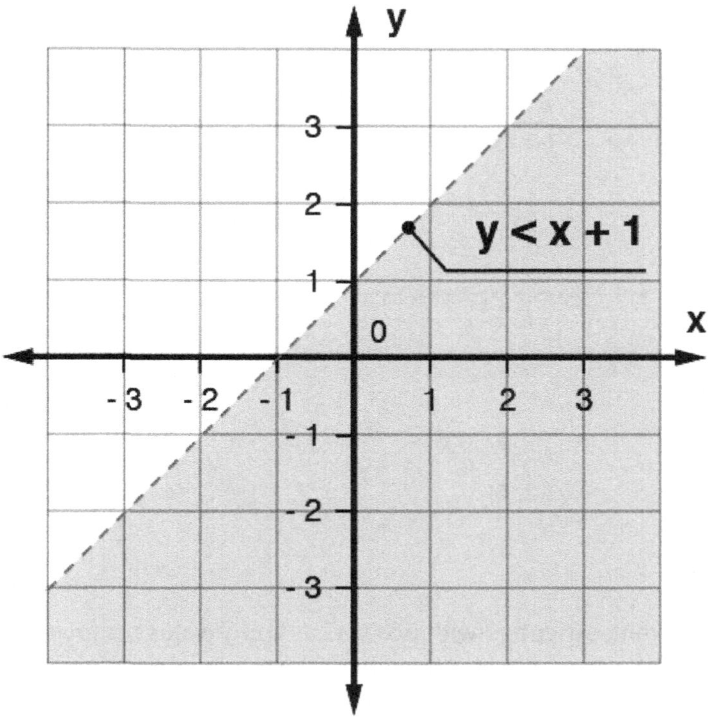

The Coordinate Plane

The coordinate plane is a way of identifying the position of a point in relation to two axes. The coordinate plane is made up of two intersecting lines, the x-axis and the y-axis. These lines intersect at a right angle, and their intersection point is called the origin. The points on the coordinate plane are labeled based on their position in relation to the origin. If a point is found 4 units to the right and 2 units up from the origin, the location is described as (4, 2). These numbers are the x- and y-coordinates, always written in the order (x, y). This point is also described as lying in the first quadrant. Every point in the first quadrant has a location that is positive in the x and y directions.

The following figure shows the coordinate plane with examples of points that lie in each quadrant.

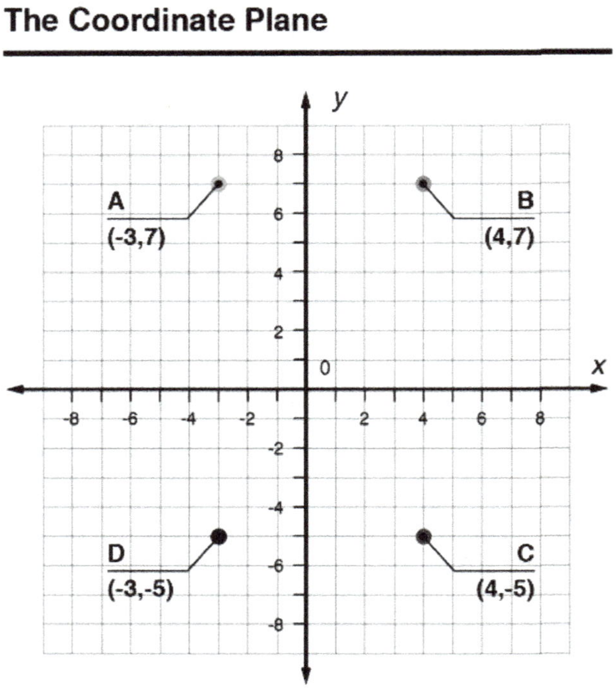

Point B lies in the first quadrant, described with positive x- and y-values, above the x-axis and to the right of the y-axis. Point A lies in the second quadrant, where the x-value is negative and the y-value is positive. This quadrant is above the x-axis and to the left of the y-axis. Point D lies in the third quadrant, where both the x- and y-values are negative. Points in this quadrant are described as being below the x-axis and to the left of the y-axis. Point C is in the fourth quadrant, where the x-value is positive and the y-value is negative.

Straight Lines

Geometric figures can be identified by matching the definition with the object. For example, a line segment is made up of two connected endpoints. A **ray** is made up of one endpoint and one extending side that goes on forever. A line has no endpoints and two sides that extend forever. These three geometric figures are shown below. What happens at A and B determines the name of each figure.

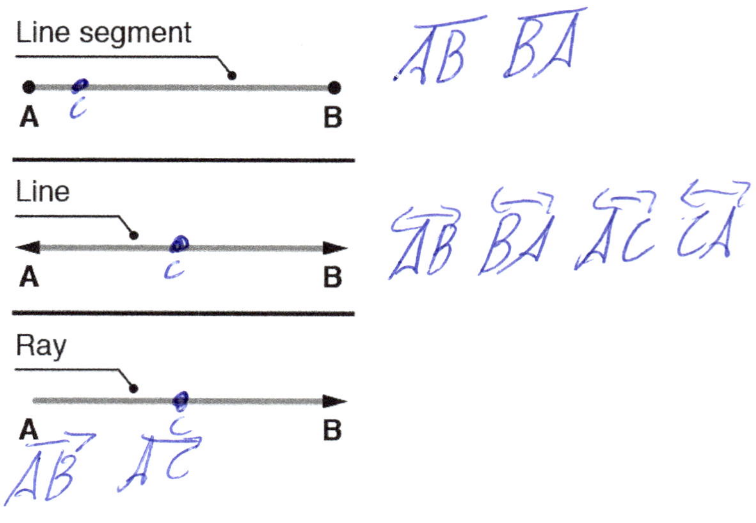

A set of lines can be parallel, perpendicular, or neither, depending on how the two lines interact. Parallel lines run alongside each other but never intersect. **Perpendicular** lines intersect at a 90-degree, or a right, angle. An example of these two sets of lines is shown below. Also shown in the figure are non-examples of these two types of lines. Because the first set of lines, in the top left corner, will eventually intersect if they continue, they are not parallel. In the second set, the lines run in the same direction and will never intersect, making them parallel. The third set, in the bottom left corner, intersect at an angle that is not right, or not 90 degrees. The fourth set is perpendicular because the lines intersect at exactly a right angle.

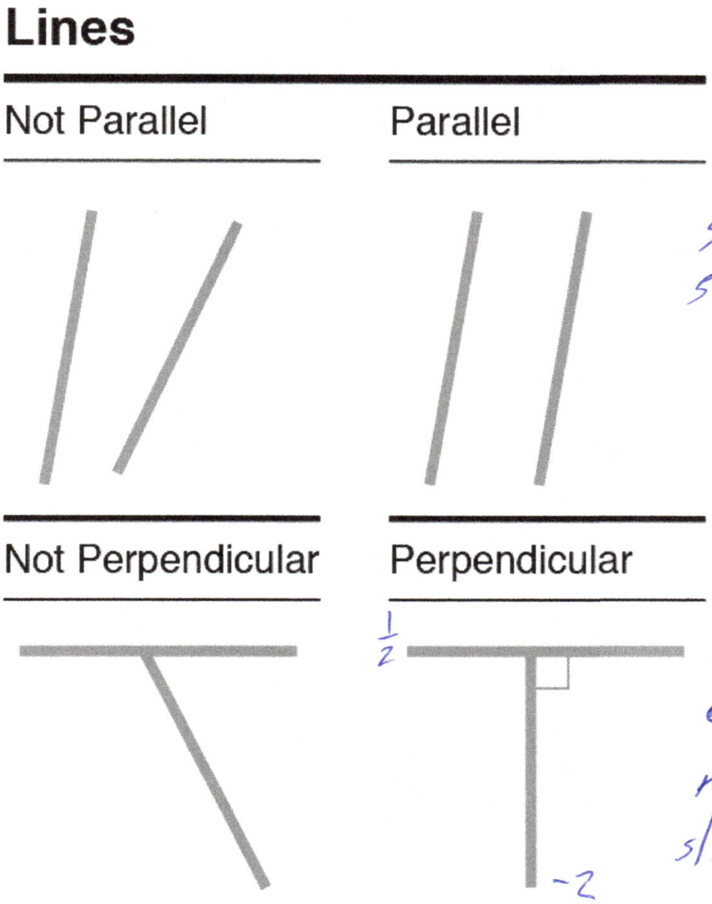

Circles in the Coordinate Plane

Recall that a **circle** is the set of all points the same distance, known as radius r, from a single point C, known as the center of the circle. The center has coordinates (h, k) and any point on the circle is an ordered pair with coordinates (x, y). A right triangle with hypotenuse r can be formed with these two points as seen here:

$(x-h)^2+(y-k)^2=r^2$

center=(h,k)

radius=r

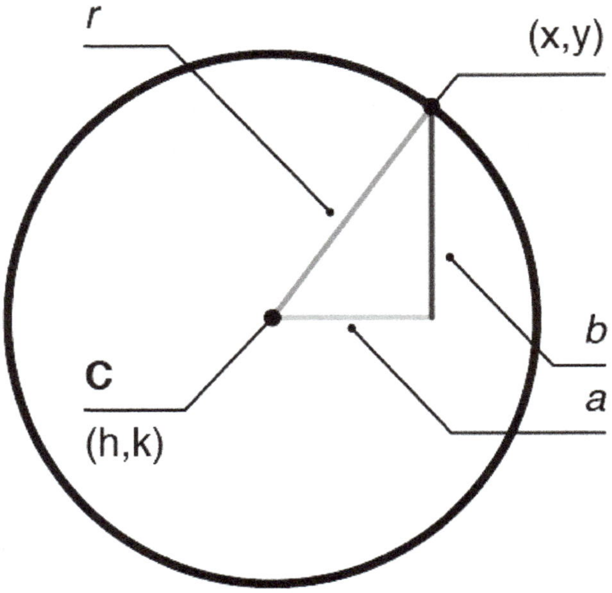

The other side lengths are a and b. The **Pythagorean Theorem** states that $a^2 + b^2 = r^2$. However, a can be replaced by $|x - h|$ and b can be replaced by $|y - k|$ because the distance between any two coordinates in the coordinate plane is the absolute value of their difference. That substitution gives $(x - h)^2 + (y - k)^2 = r^2$, which is the formula used to find the equation of any circle with center (h, k) and radius r. Therefore, if any problem gives the coordinates of the center of a circle and its radius length, this is the equation in two variables that allows any other point on the circle to be found.

Oftentimes, the center or the radius of a circle are not easily seen in the given equation of the circle. If the equation is in standard form of a polynomial equation like $ax^2 + ay^2 + cx + dy + e = 0$, the algebraic technique of completing the square must be used to find the coordinates of the center and the radius. Completing the square must be done within both variables x and y. First, the constant term needs to be subtracted off of both sides of the equation, and then the x and y terms need to be grouped together. Then, the entire equation needs to be divided by a. Then, divide the coefficient of the x term by 2, square it, and add that value to both sides of the equation. This value should be grouped with the x terms. Next, divide the coefficient of the y term by 2, square it, and add it to both sides of the equation, grouping it with the y terms. The trinomial in both x and y are now perfect square trinomials and can be factored into squares of a binomial. This process results in $(x - h)^2 + (y - k)^2 = r^2$, showing the radius and coordinates of the center.

Transformations and Symmetry

Two-dimensional figures can undergo various types of transformations in the plane. They can be shifted horizontally and vertically, reflected, compressed, or stretched.

A shift, also known as a slide or a translation, moves the shape in one direction. Here is a **picture of a shift:**

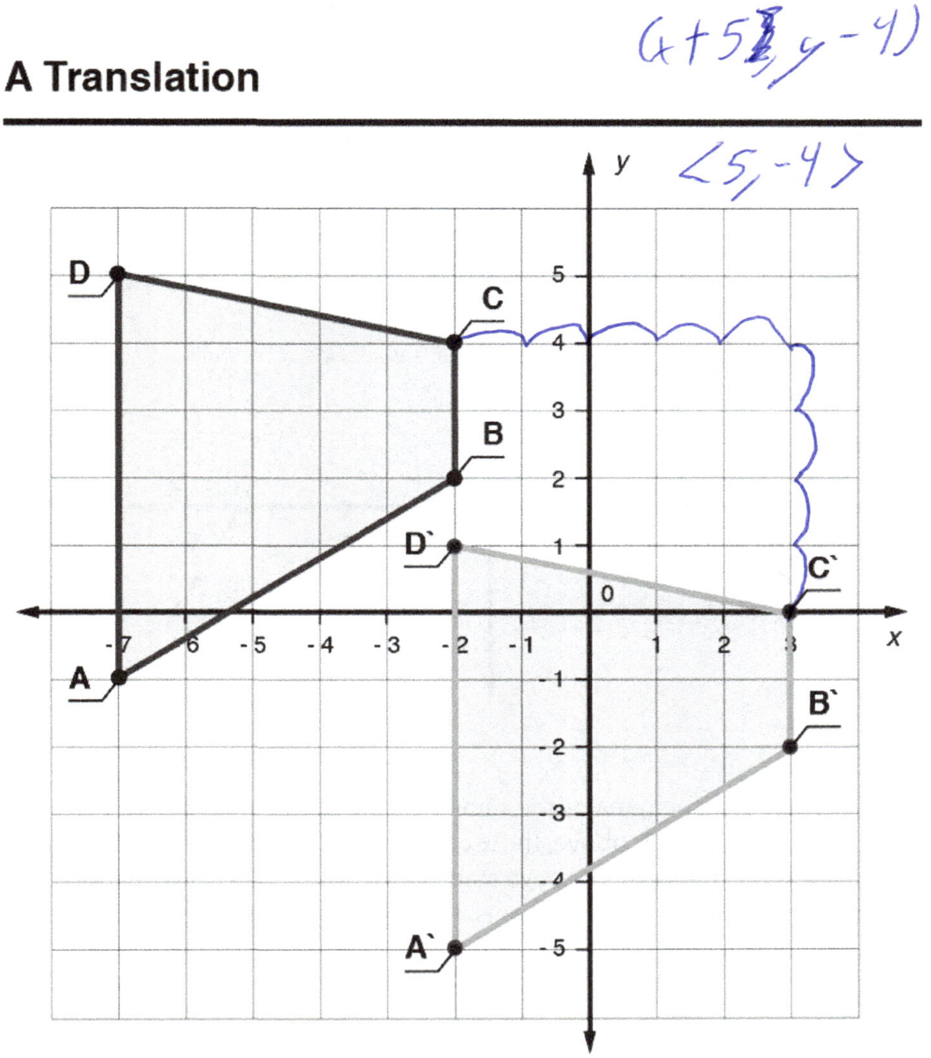

Notice that the size of the original shape has not changed at all. If the shift occurs within a **Cartesian coordinate system**, the standard *x* and *y* coordinate plane, it can be represented by adding to or subtracting from the *x* and *y* coordinates of the original shape. All vertices will move the same number of units because the shape and size of the shape do not change.

A figure can also be reflected, or flipped, over a given line known as the **line of reflection**. For example, consider the following picture:

A Reflection Over the Y-Axis

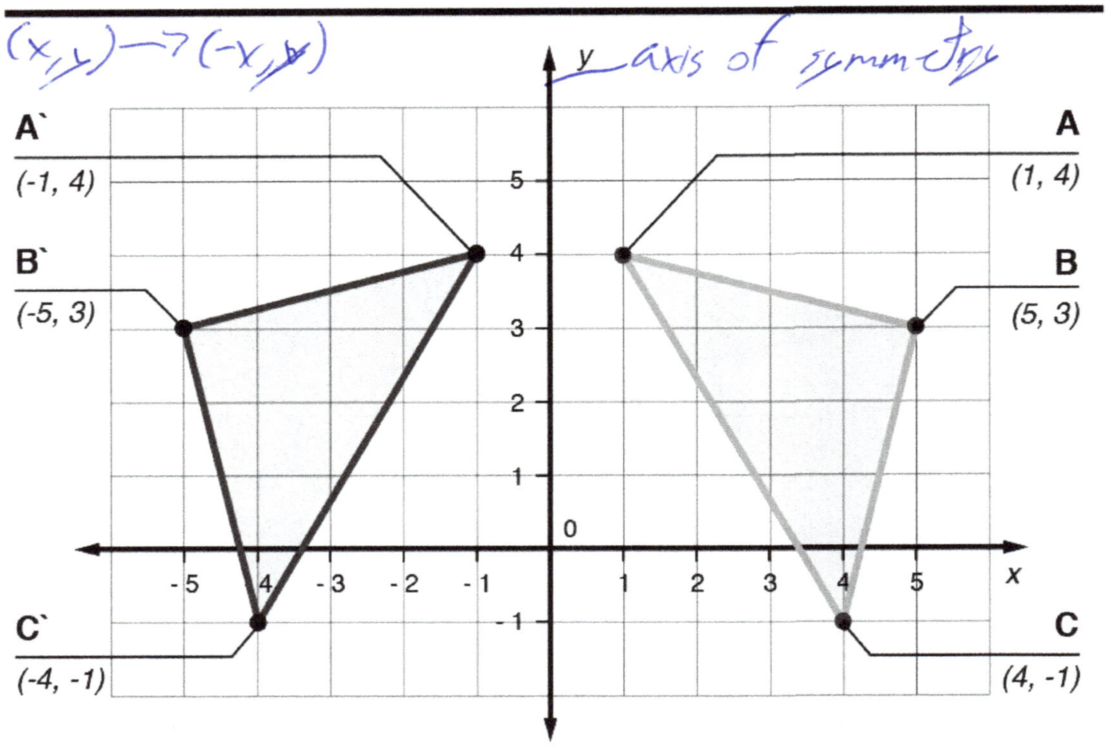

After the reflection, the original shape remains the same size, but the coordinates change. For example, if a shape gets reflected over the *y*-axis, as above, the *x*-coordinate stays the same, but the *y*-coordinates are made negative. For example, the triangle above starts in the first and second quadrants, but it is reflected over *y*-axis to the third and fourth quadrants. Point A has the initial coordinates of (1, 4), but in the reflection, the point A' becomes (-1, 4).

Similarly, if the shape is reflected over the x-axis, the y-coordinate stays the same, but the x-coordinates are made negative. For instance, in the graphic below, the point C at (3, 5) becomes C' at (3, -5).

A Reflection Over the X-Axis

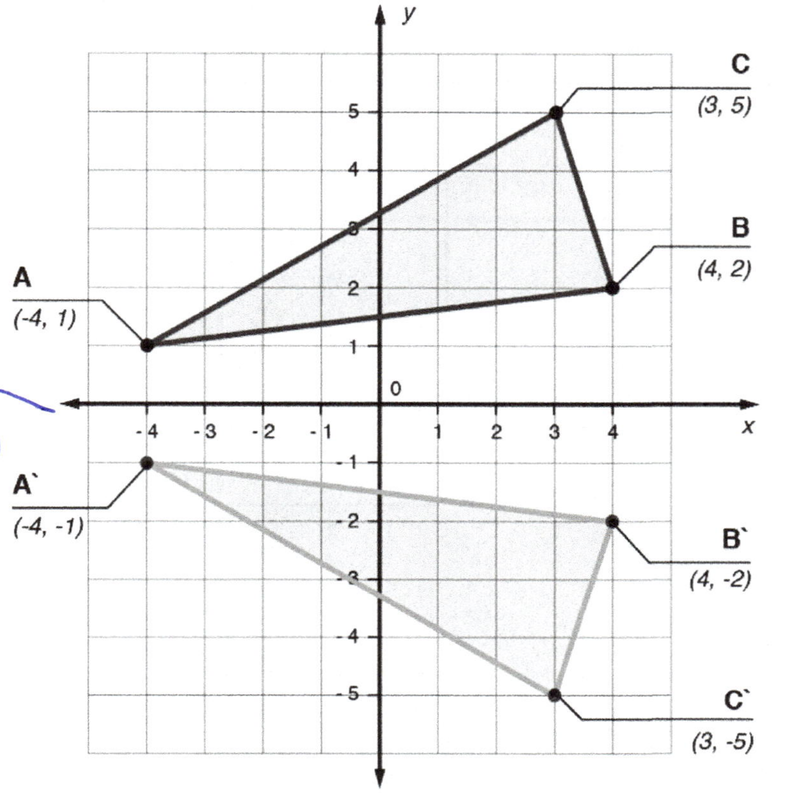

A compression or stretch of a figure involves changing the size of the original figure; both transformations are called **dilations**. A compression shrinks the size of the figure. We can think about this as a multiplication process by multiplying times a value between 0 and 1. A stretch of a figure results in a figure larger than the original shape. If we consider multiplication, the factor would be greater than 1. Here are is picture of a dilation that is comprised of a stretch in which the original square doubled in size:

A Dilation with a Scale Factor of 2

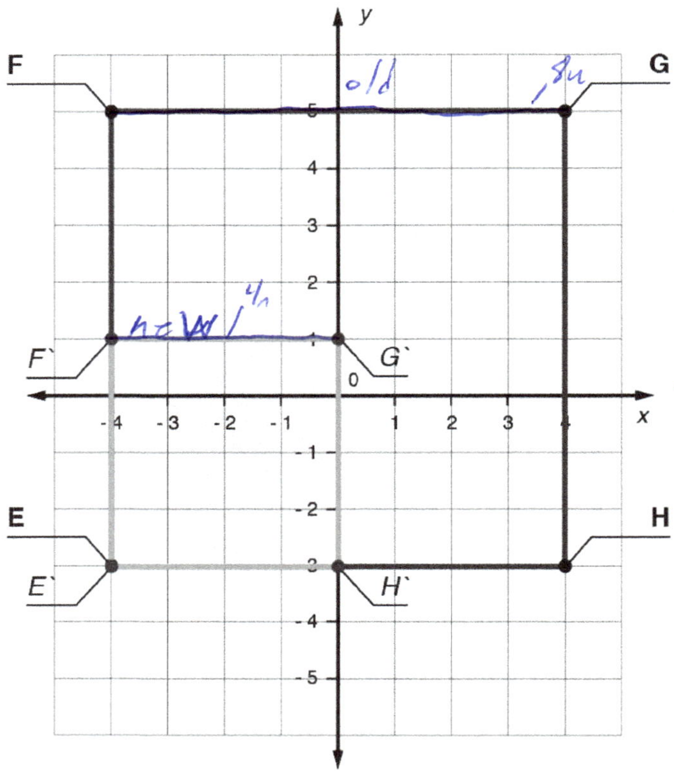

If a shape within the Cartesian coordinate system gets stretched, its coordinates get multiplied by a number greater than 1, and if a shape gets compressed, its coordinates get multiplied by a number between 0 and 1.

A figure can undergo any combination of transformations. For instance, it can be shifted, reflected, and stretched at the same time.

Two figures are **congruent** if they have the same shape and same size, meaning same angle measurements and equal side lengths. Two figures are similar if they have the same angle measurements but not side lengths. Basically, angles are congruent in similar triangles and their side lengths are constant multiples of each other. Proving two shapes are similar involves showing that all angles are the same; proving two shapes are congruent involves showing that all angles are the same *and* that all sides are the same. If two pairs of angles are congruent in two triangles, then those triangles are similar because their third angles have to be equal due to the fact that all three angles add up to 180 degrees.

There are five main theorems that are used to prove congruence in triangles. Each theorem involves showing that different combinations of sides and angles are the same in two triangles, which proves the triangles are congruent. The **side-side-side (SSS) theorem** states that if all sides are equal in two triangles, the triangles are congruent. The **side-angle-side (SAS) theorem** states that if two pairs of sides and the included angles are equal in two triangles then the triangles are congruent. Similarly, the **angle-side-angle (ASA) theorem** states that if two pairs of angles and the included side lengths are equal in two triangles, the triangles are similar. The **angle-angle-side (AAS) theorem** states that two triangles are congruent if they have two pairs of congruent angles and a pair of corresponding equal side lengths that are not included. Finally, the **hypotenuse-leg (HL) theorem** states that if two right triangles have equal hypotenuses and an equal pair of shorter sides, the triangles are congruent. An important item to note is that angle-angle-angle (AAA) is not enough information to prove congruence because the three angles could be equal in two triangles, but their sides could be different lengths.

Sets of Points in the Plane

Shapes can be plotted on the coordinate plane to identify the location of each vertex and the length of each side. The original triangle (called the **preimage**) is seen in the figure below in the first quadrant. The reflection of this triangle is in the second quadrant. A reflection across the y-axis can be found by determining each point's distance to the y-axis and moving it that same distance on the opposite side. For example, the point C is located at (4, 1). The reflection of this point moves to (-4, 1) when reflected across the y-axis. The original point A is located at (1, 3), and the reflection across the y-axis is located at (-1, 5). It is evident that the reflection across the y-axis changes the sign of the x-coordinate. A reflection across the x-axis changes the sign of the y-coordinate instead.

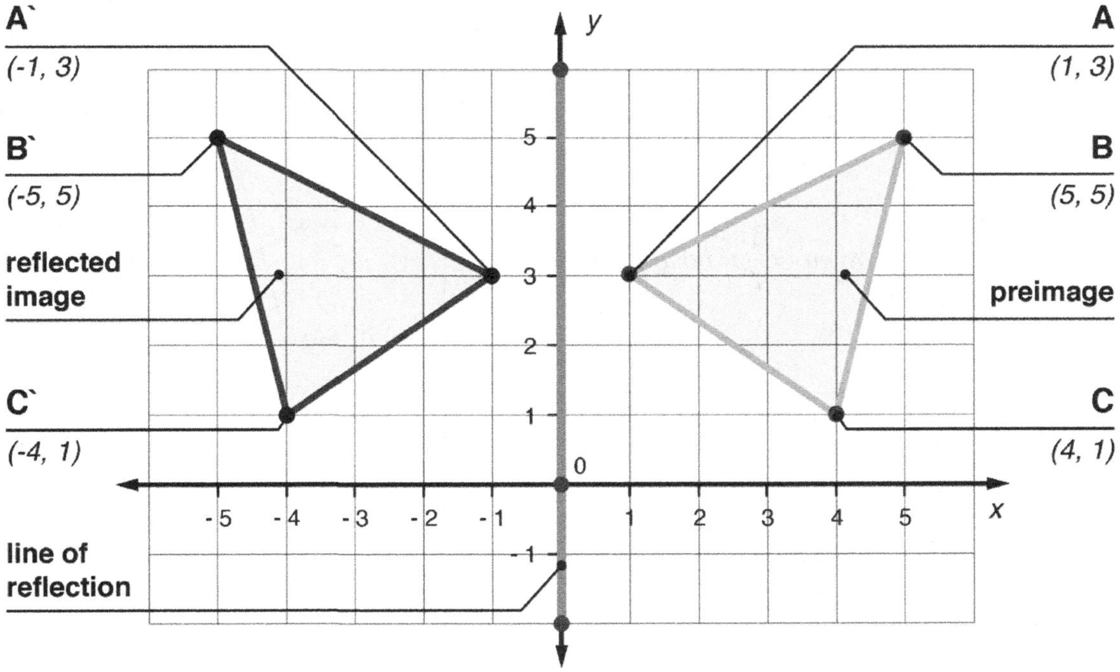

Linear, Area, and Three-Dimensional Measurements

Perimeter

Perimeter is the distance around an object. The perimeter of an object can be found by adding the lengths of all sides. Perimeter may be used in problems dealing with lengths around objects such as fences or borders. It may also be used in finding missing lengths or working backwards. If the perimeter is given, but a length is missing, use subtraction to find the missing length. Given a square with side length s, the formula for perimeter is $P = 4s$. Given a rectangle with length l and width w, the formula for perimeter is $P = 2l + 2w$. The perimeter of a triangle is found by adding the three side lengths, and the perimeter of a trapezoid is found by adding the four side lengths. The units for perimeter are always the original units of length, such as meters, inches, miles, etc. When discussing a circle, the distance around the object is referred to as its circumference, not perimeter. The formula for the circumference of a circle is $C = 2\pi r$, where r represents the radius of the circle. This formula can also be written as $C = d\pi$, where d represents the diameter of the circle.

Area

Area is the two-dimensional space covered by an object. These problems may include the area of a rectangle, a yard, or a wall to be painted. Finding the area may require a simple formula or multiple formulas used together. The units for area are square units, such as square meters, square inches, and square miles. Given a square with side length s, the formula for its area is $A = s^2$.

Some other formulas for common shapes are shown below.

Shape	**Formula**	**Graphic**
Rectangle	$Area = length \times width$	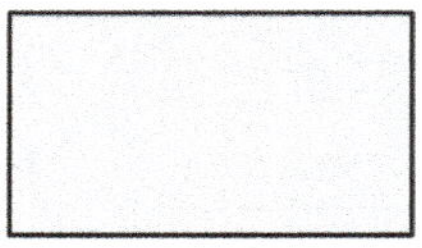
Triangle	$Area = \frac{1}{2} \times base \times height$	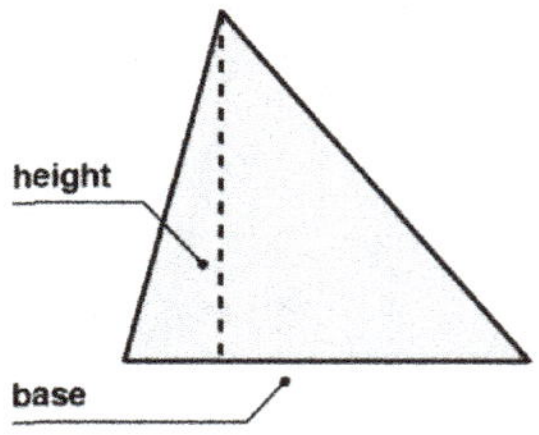
Circle	$Area = \pi \times radius^2$	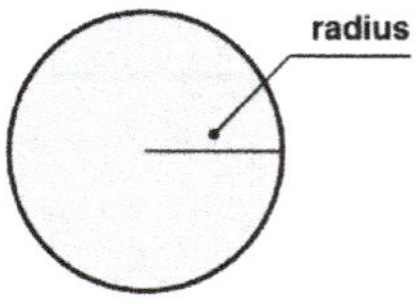

The following formula, not as widely used as those shown above, but very important, is the area of a trapezoid:

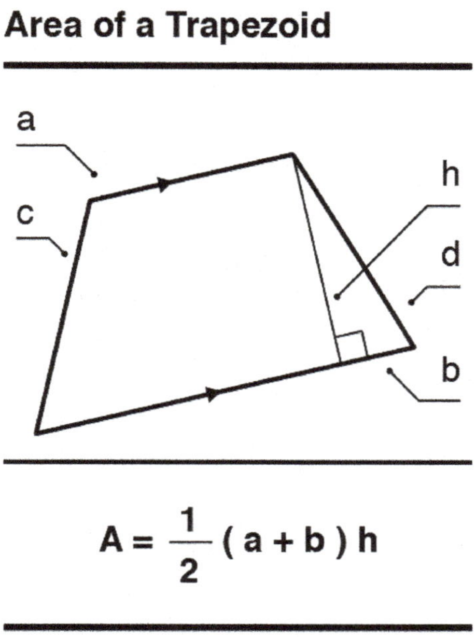

$A = \frac{1}{2}(b_1 + b_2)h$

To find the area of the shapes above, use the given dimensions of the shape in the formula. Complex shapes might require more than one formula. To find the area of the figure below, break the figure into two shapes. The rectangle has dimensions 11 cm by 6 cm. The triangle has dimensions 3 cm by 6 cm. Plug the dimensions into the rectangle formula: $A = 11 \times 6$. Multiplication yields an area of 66 cm^2. The triangle's area can be found using the formula:

$$A = \frac{1}{2} \times 4 \times 6$$

Multiplication yields an area of 12 cm^2. Add the two areas to find the total area area of the figure, which is 78 cm^2.

$\frac{1}{2}(11+15)6$

$\frac{1}{2}(26)6$

$13(6)$

$A = 18 cm^2$

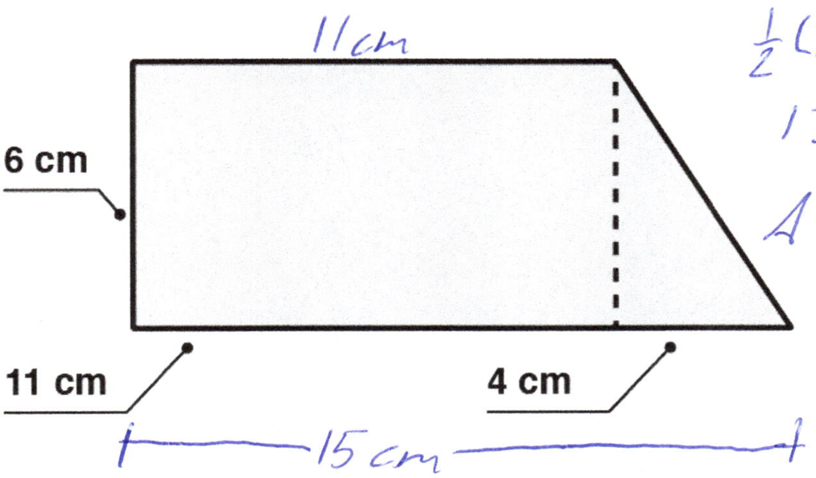

Instead of combining areas, some problems may require subtracting them, or finding the difference.

To find the area of the shaded region in the figure below, determine the area of the whole figure. Then subtract the area of the circle from the whole.

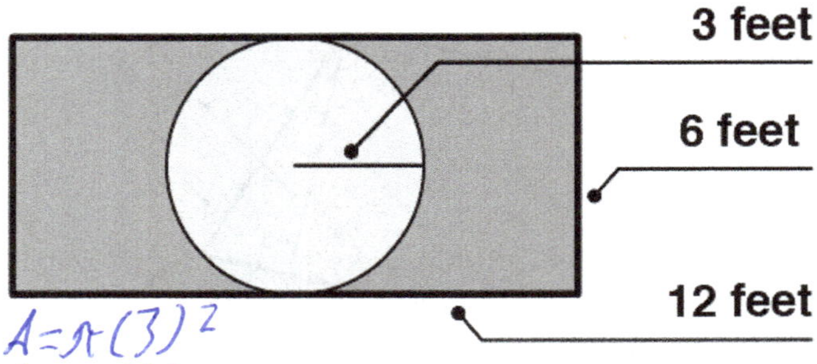

The following formula shows the area of the outside rectangle:

$$A = 12 \times 6 = 72 \text{ ft}^2$$

The area of the inside circle can be found by the following formula:

$$A = \pi(3)^2 = 9\pi = 28.3 \text{ ft}^2$$

As the shaded area is outside the circle, the area for the circle can be subtracted from the area of the rectangle to yield an area of 43.7 ft^2.

Volume

Volume is three-dimensional and describes the amount of space that an object occupies, but it's different from area because it has three dimensions instead of two. The units for volume are cubic units, such as cubic meters, cubic inches, and cubic miles. Volume can be found by using formulas for common objects such as cylinders and boxes.

The following chart shows a diagram and formula for the volume of two objects.

Shape	Formula	Diagram
Rectangular Prism (box)	$V = length \times width \times height$	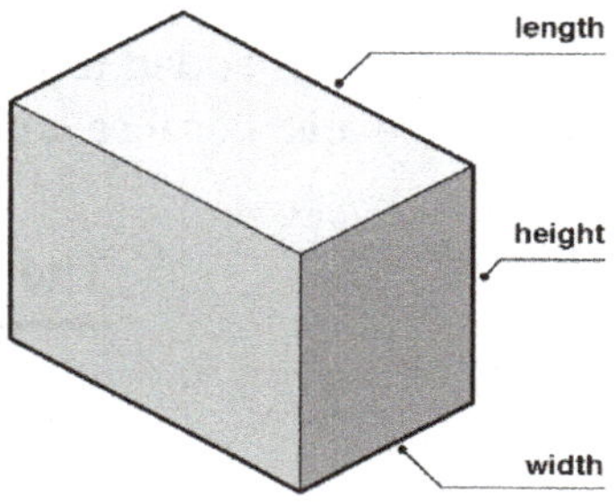
Cylinder	$V = \pi \times radius^2 \times height$	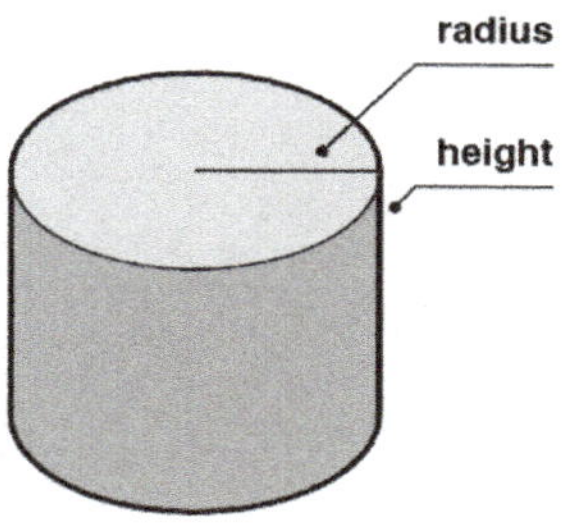

Volume formulas of these two objects are derived by finding the area of the bottom two-dimensional shape, such as the circle or rectangle, and then multiplying times the height of the three-dimensional shape. Other volume formulas include the volume of a cube with side length s: $V = s^3$; the volume of a sphere with radius r: $V = \frac{4}{3}\pi r^3$; and the volume of a cone with radius r and height h:

$$V = \frac{1}{3}\pi r^2 h$$

If a soda can has a height of 5 inches and a radius on the top of 1.5 inches, the volume can be found using one of the given formulas. A soda can is a cylinder. Knowing the given dimensions, the formula can be completed as follows:

$$V = \pi(radius)^2 \times height$$

$$\pi(1.5 \text{ in})^2 \times 5 \text{ in} = 35.325 \text{ in}^3$$

Notice that the units for volume are inches cubed because it refers to the number of cubic inches required to fill the can.

Surface Area

Surface area is defined as the area of the surface of a figure. A **pyramid** has a surface made up of four triangles and one square. To calculate the surface area of a pyramid, the areas of each individual shape are calculated. Then the areas are added together. This method of decomposing the shape into two-dimensional figures to find area, then adding the areas, can be used to find surface area for any figure. Once these measurements are found, the area is described with square units. For example, the following figure shows a rectangular prism. The figure beside it shows the rectangular prism broken down into two-dimensional shapes, or rectangles. The area of each rectangle can be calculated by multiplying the length by the width. The area for the six rectangles can be represented by the following expression:

$$5 \times 6 + 5 \times 10 + 5 \times 6 + 6 \times 10 + 5 \times 10 + 6 \times 10$$

The total for all these areas added together is $280\ \text{m}^2$, or 280 square meters. This measurement represents the surface area because it is the area of all six surfaces of the rectangular prism.

The Net of a Rectangular Prism

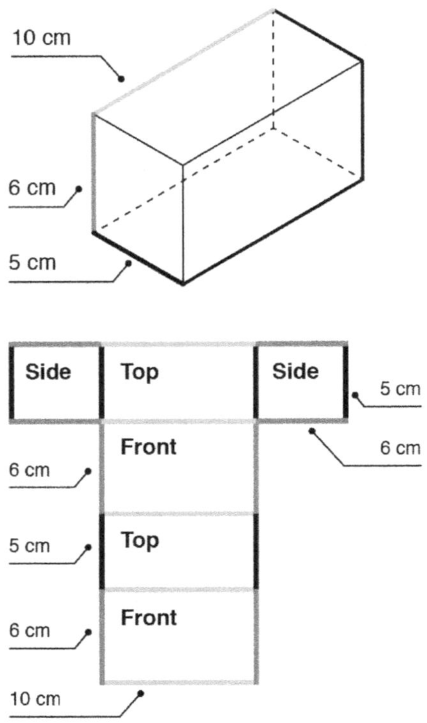

Another shape that has a surface area is a cylinder. The shapes used to make up the **cylinder** are two circles and a rectangle wrapped around between the two circles. A common example of a cylinder is a can. The two circles that make up the bases are obvious shapes. The rectangle can be more difficult to see, but the label on a can will help illustrate it. When the label is removed from a can and laid flat, the shape is a rectangle. When the areas for each shape are needed, there will be two formulas. The first is the area for the circles on the bases. This area is given by the formula $A = \pi r^2$. There will be two of these areas—one for the top and one for the bottom if the can (cylinder) is standing upright on a shelf. Then the area of the rectangle must be determined. The width of the rectangle is equal to the height of the can, *h*. The length of the rectangle is equal to the circumference of the base circle, $2\pi r$. The area for the rectangle can be found by using the formula $A = 2\pi r \times \text{h}$. By adding the two areas for the bases and the area of the rectangle, the surface area of the cylinder can be found, described in units squared.

Solving for Missing Values in Triangles, Circles, and Other Figures

Solving for missing values in shapes requires knowledge of the shape and its characteristics. For example, a triangle has three sides and three angles that add up to 180 degrees. If two angle measurements are given, the third can be calculated. For the triangle below, the one given angle has a measure of 55 degrees. The missing angle is x. The third angle is labeled with a square, which indicates a measure of 90 degrees. Because all angles must add up to 180 degrees,, the following equation can be used to find the missing x-value:

$$55° + 90° + x = 180°$$

Adding the two given angles and subtracting the total from 180 gives an answer of 35 degrees.

180-90-55=x

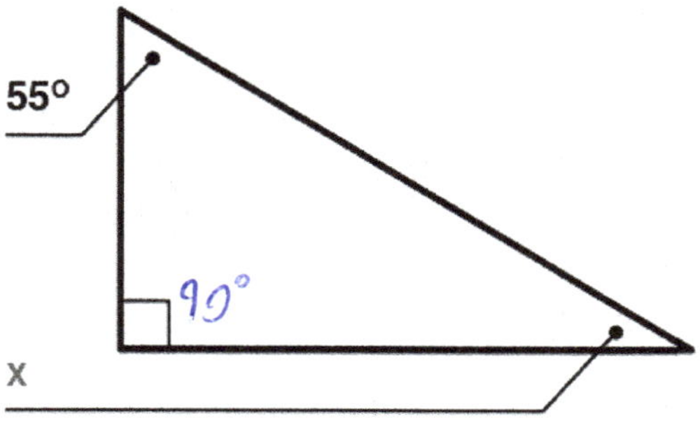

A similar problem can be solved with circles. If the radius is given but the circumference is unknown, the circumference can be calculated based on the formula $C = 2\pi r$. This example can be used in the figure below. The radius can be substituted for r in the formula. Then the circumference can be found as:

$$C = 2\pi \times 8 = 16\pi = 50.24 \text{ cm}$$

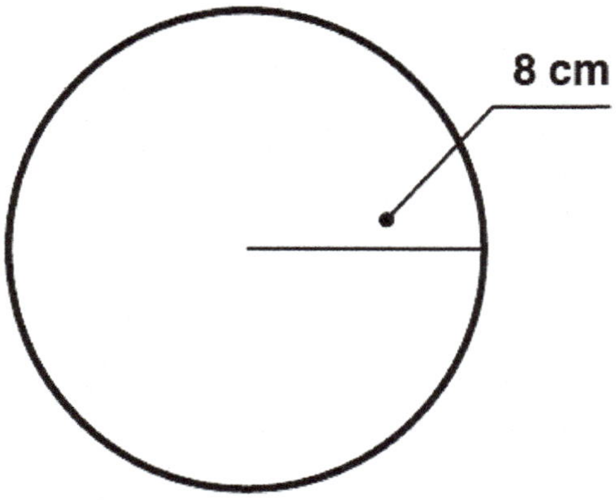

Other figures that may have missing values could be the length of a square, given the area, or the perimeter of a rectangle, given the length and width. All of the missing values can be found by first identifying all the characteristics that are known about the shape, then looking for ways to connect the missing value to the given information.

Probabilistic and Statistical Reasoning

Interpreting Categorical and Quantitative Data

They all organize, categorize, and compare data, and they come in different shapes and sizes. Each type has its own way of showing information, whether through a column, shape, or picture. To answer a question relating to a table, chart, or graph, some steps should be followed. First, the problem should be read thoroughly to determine what is being asked to determine what quantity is unknown. Then, the title of the table, chart, or graph should be read. The title should clarify what data is actually being summarized in the table. Next, look at the key and labels for both the horizontal and vertical axes, if they are given. These items will provide information about how the data is organized. Finally, look to see if there is any more labeling inside the table. Taking the time to get a good idea of what the table is summarizing will be helpful as it is used to interpret information.

Tables are a good way of showing a lot of information in a small space. The information in a table is organized in columns and rows. For example, a table may be used to show the number of votes each candidate received in an election. By interpreting the table, one may observe which candidate won the election and which candidates came in second and third. In using a bar chart to display monthly rainfall amounts in different countries, rainfall can be compared between countries at different times of the year. Graphs are also a useful way to show change in variables over time, as in a line graph, or percentages of a whole, as in a pie graph.

The table below relates the number of items to the total cost. The table shows that one item costs $5. By looking at the table further, five items cost $25, ten items cost $50, and fifty items cost $250. This cost can be extended for any number of items. Since one item costs $5, then two items would cost $10. Though this information is not in the table, the given price can be used to calculate unknown information.

Number of Items	1	5	10	50
Cost ($)	5	25	50	250

A **bar graph** is a graph that summarizes data using bars of different heights. It is useful when comparing two or more items or when seeing how a quantity changes over time. It has both a horizontal and vertical axis. To interpret bar graphs, recognize what each bar represents and connect that to the two variables. The bar graph below shows the scores for six people during three different games. The different colors of the bars distinguish between the three games, and the height of the bar indicates their score for that game. William scored 25 on game 3, and Abigail scored 38 on game 3. By comparing the bars, it is obvious that Williams scored lower than Abigail.

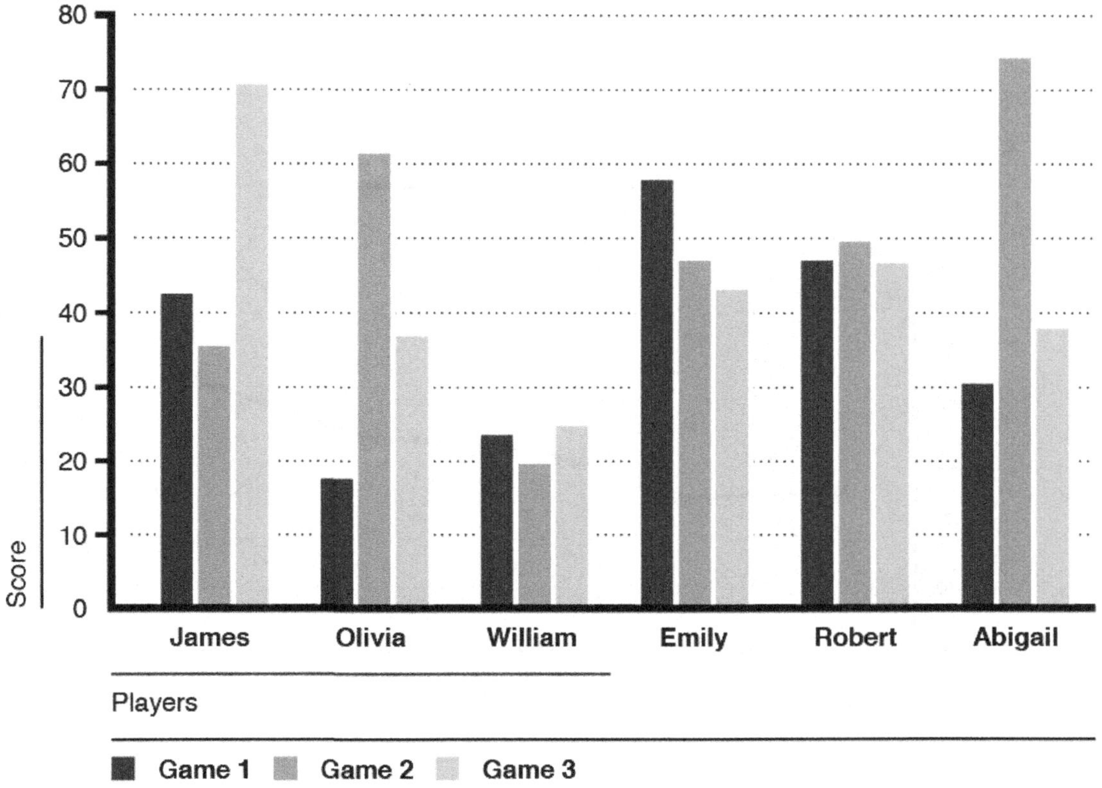

A line graph is a way to compare two variables that are plotted on opposite axes of a graph. The line indicates a continuous change as it rises or falls. The line's rate of change is known as its slope. The horizontal axis often represents a variable of time. Readers can quickly see if an amount has grown or decreased over time. The bottom of the graph, or the x-axis, shows the units for time, such as days, hours, months, etc. If there are multiple lines, a comparison can be made between what the two lines represent. For example, the following line graph, shown previously, displays the change in temperature over five days. The top line represents the high, and the bottom line represents the low for each day. Looking at the top line alone, the high decreases for a day, then increases on Wednesday. Then it decreases on Thursday and increases again on Friday. The low temperatures have a similar trend, shown in the bottom line. The range in temperatures each day can also be calculated by finding the difference between the top line and bottom line on a particular day. On Wednesday, the range was 14 degrees, from 62 to 76° F.

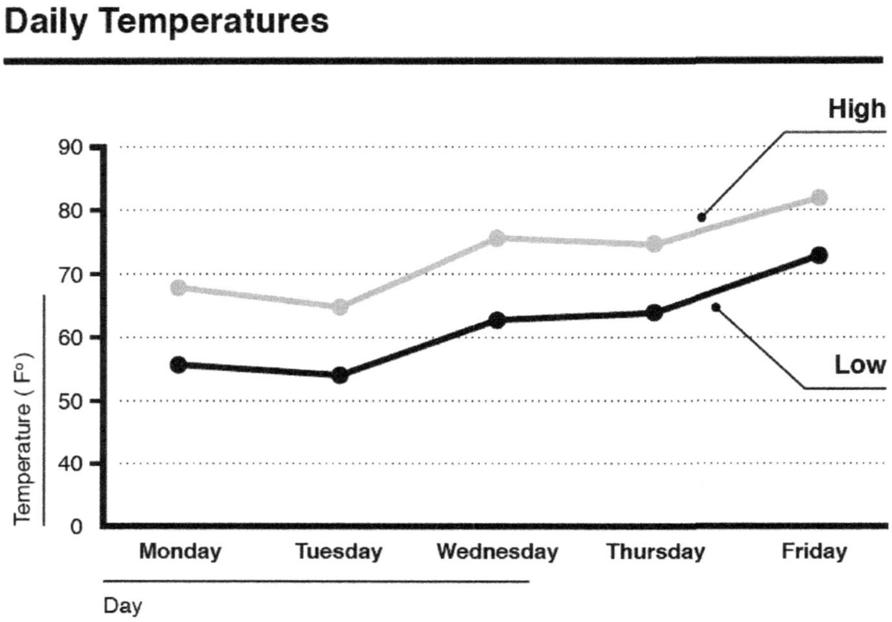

Pie charts show percentages of a whole; they are circular representations of data used to highlight numerical proportions. Each category represents a piece of the pie, and together, all of the pieces make up a whole. The size of each pie slice is proportional to the amount it represents; therefore, a reader can quickly make comparisons by visualizing the sizes of the pieces. They can be useful for comparison between different categories. The following pie chart is a simple example of three different categories shown in comparison to each other.

Light gray represents cats, dark gray represents dogs, and the medium shade of gray represents other pets. These three equal pieces each represent just more than 33 percent, or $\frac{1}{3}$ of the whole. Values 1 and 2 may be combined to represent $\frac{2}{3}$ of the whole. In an example where the total pie represents 75,000 animals, then cats would be equal to $\frac{1}{3}$ of the total, or 25,000. Dogs would equal 25,000 and other pets also equal 25,000.

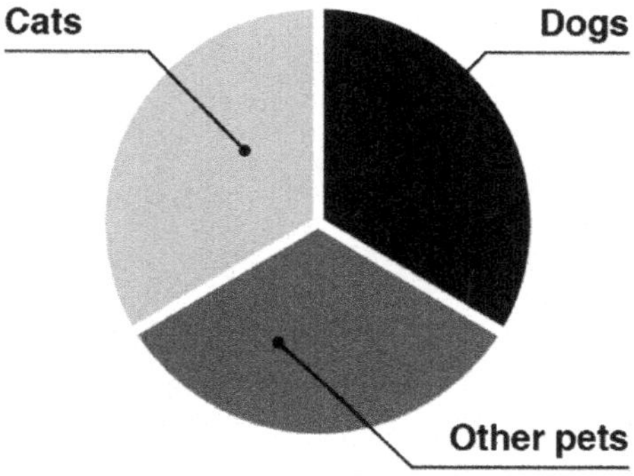

The fact that a circle is 360 degrees is used to create a pie chart. Because each piece of the pie is a percentage of a whole, that percentage is multiplied times 360 to get the number of degrees each piece represents. In the example above, each piece is $\frac{1}{3}$ of the whole, so each piece is equivalent to 120 degrees. Together, all three pieces add up to 360 degrees.

Stacked bar graphs are also used fairly frequently when comparing multiple variables at one time. They combine some elements of both pie charts and bar graphs, using the organization of bar graphs and the proportionality aspect of pie charts. The following is an example of a stacked bar graph that represents the number of students in a band playing drums, flute, trombone, and clarinet. Each bar graph is broken up further into girls and boys.

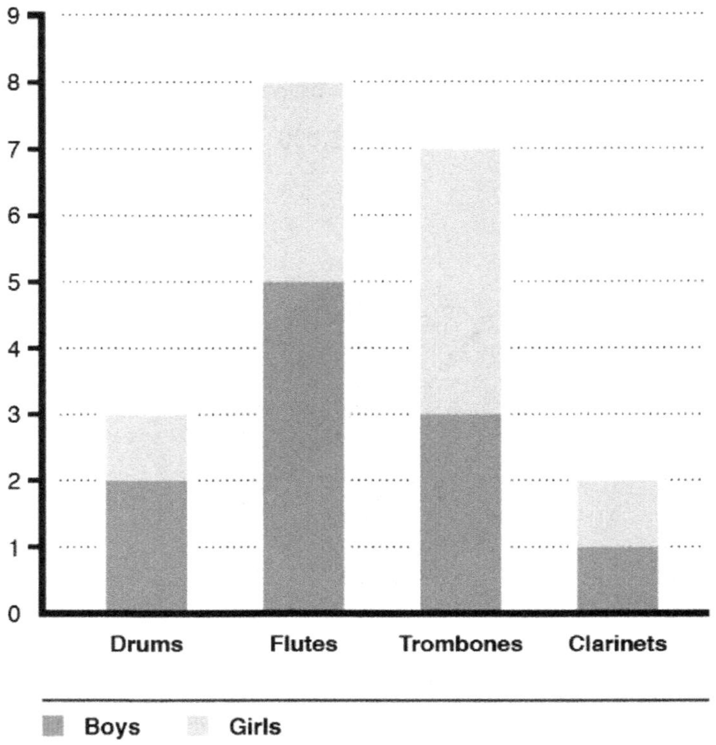

To determine how many boys play trombone, refer to the darker portion of the trombone bar, which indicates three boys.

As mentioned, a **scatterplot** is another way to represent paired data. It uses Cartesian coordinates, like a line graph, meaning it has both a horizontal and vertical axis. Each data point is represented as a dot on the graph. The dots are never connected with a line. For example, the following is a scatterplot showing the connection between people's ages and heights.

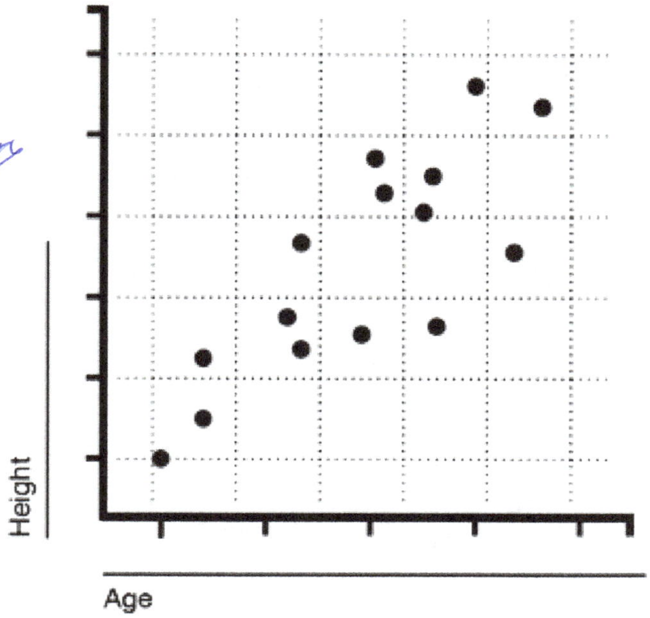

A scatterplot, also known as a **scattergram,** can be used to predict another value and to see if a correlation exists between two variables in a set of data. If the data resembles a straight line, then it is associated, or correlated. The following is an example of a scatterplot in which the data does not seem to have an association:

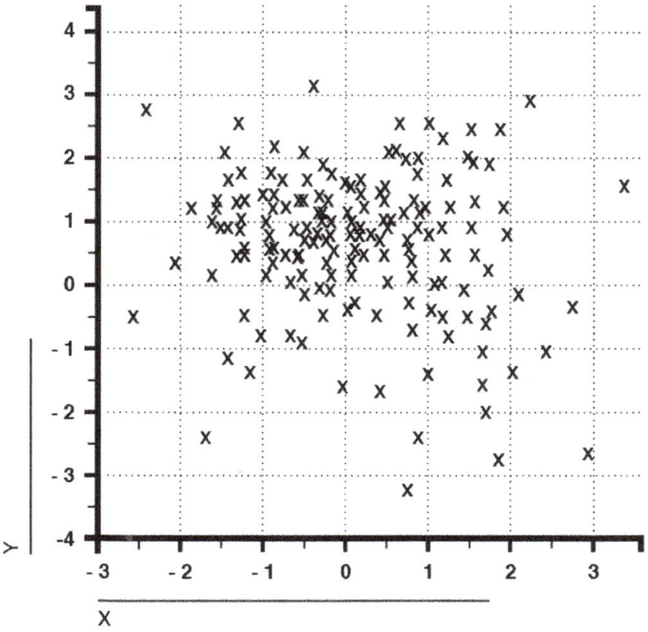

Sets of numbers and other similarly organized data can also be represented graphically. Venn diagrams are a common way to do so. A **Venn diagram** represents each set of data as a circle. The circles overlap, showing that each set of data is overlapping. A Venn diagram is also known as a **logic diagram** because it visualizes all possible logical combinations between two sets. Common elements of two sets are represented by the area of overlap. The following is an example of a Venn diagram of two sets A and B:

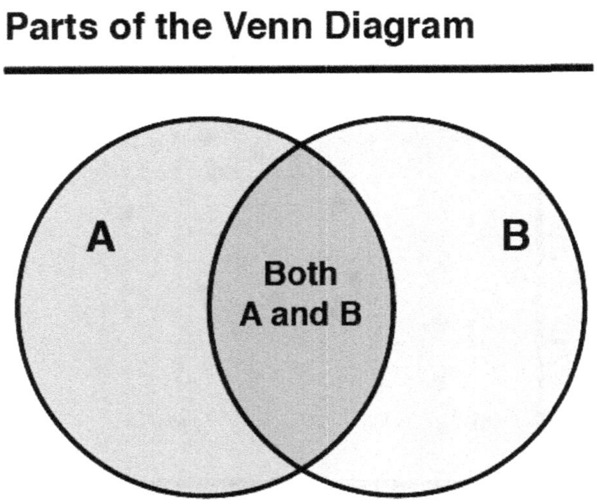

Another name for the area of overlap is the **intersection.** The intersection of A and B, written $A \cap B$, contains all elements that are in both sets A and B. The **union** of A and B, $A \cup B$, contains all elements in both sets A and B. Finally, the **complement** of $A \cup B$ is equal to all elements that are not in either set A or set B. These elements are placed outside of the circles.

The following is an example of a Venn diagram representing 24 students who were surveyed about their siblings. Ten students only had a brother, seven students only had a sister, and five had both a brother and a sister. Therefore, five is the intersection, represented by the section where the circles overlap. Two students did not have a brother or a sister. Therefore, two is the complement and is placed outside of the circles.

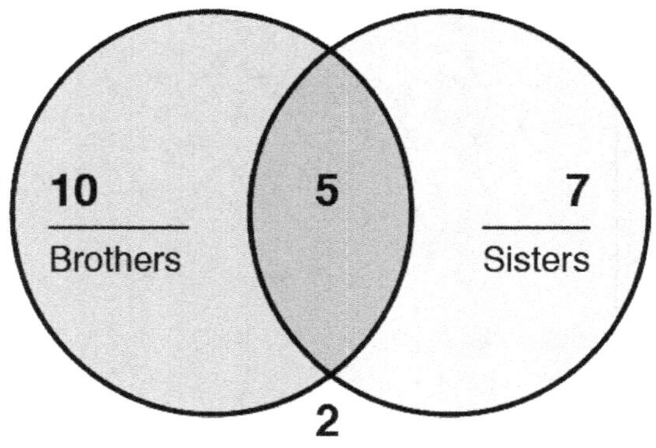

Venn diagrams can have more than two sets of data. The more circles, the more logical combinations are represented by the overlapping. The following is a Venn diagram that represents sock colors worn by a class of students. There were 30 students surveyed. The innermost region represents those students that had green, pink, and blue on their socks (perhaps in a striped pattern). Therefore, two students had all three colors.. In this example, all students had at least one color on their socks, so there is no complement.

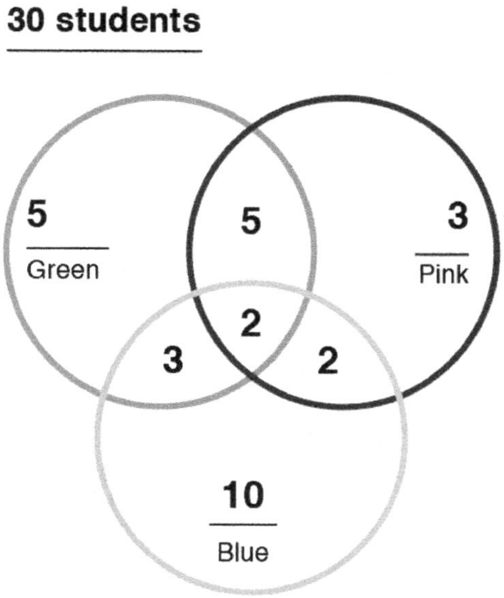

Venn diagrams are typically not drawn to scale; however, if they are, and if each circle's area is proportional to the amount of data it represents, then it is called an area-proportional Venn diagram.

Describing Distributions

One way information can be interpreted from tables, charts, and graphs is through statistics. The three most common calculations for a set of data are the mean, median, and mode. These three are called **measures of central tendency**, which are helpful in comparing two or more different sets of data. The **mean** refers to the average and is found by adding up all values and dividing the total by the number of values. In other words, the mean is equal to the sum of all values divided by the number of data entries. For example, if you bowled a total of 532 points in 4 bowling games, your mean score was $\frac{532}{4} = 133$ points per game. Students can apply the concept of mean to calculate what score they need on a final exam to earn a desired grade in a class.

The **median** is found by lining up values from least to greatest and choosing the middle value. If there is an even number of values, then calculate the mean of the two middle amounts to find the median. For example, the median of the set of dollar amounts $5, $6, $9, $12, and $13 is $9. The median of the set of dollar amounts $1, $5, $6, $8, $9, $10 is $7, which is the mean of $6 and $8. The **mode** is the value that occurs the most. The mode of the data set {1, 3, 1, 5, 5, 8, 10} actually refers to two numbers: 1 and 5. In this case, the data set is **bimodal** because it has two modes. A data set can have no mode if no amount is repeated. Another useful statistic is range. The **range** for a set of data refers to the difference between the highest and lowest value.

In some cases, numbers in a list of data might have weights attached to them. In that case, a **weighted mean** can be calculated. A common application of a weighted mean is GPA. In a semester, each class is assigned a number of credit hours, its weight, and at the end of the semester each student receives a grade. To compute GPA, an A is a 4, a B is a 3, a C is a 2, a D is a 1, and an F is a 0. Consider a student that takes a 4-hour English class, a 3-hour math class, and a 4-hour history class and receives all B's. The weighted mean, GPA, is found by multiplying each grade times its weight, number of credit hours, and dividing by the total number of credit hours. Therefore, the student's GPA is:

$$\frac{3 \times 4 + 3 \times 3 + 3 \times 4}{11} = \frac{33}{11} = 3.0.$$

The following bar chart shows how many students attend a cycle class on each day of the week. To find the mean attendance for the week, add each day's attendance together:

$$10 + 7 + 6 + 9 + 8 + 14 + 4 = 58$$

Then divide the total by the number of days:

$$58 \div 7 = 8.3.$$

The mean attendance for the week was 8.3 people. The median attendance can be found by putting the attendance numbers in order from least to greatest: 4, 6, 7, 8, 9, 10, 14, and choosing the middle number: 8 people. This set of data has no mode because no numbers repeat. The range is 10, which is found by finding the difference between the lowest number, 4, and the highest number, 14.

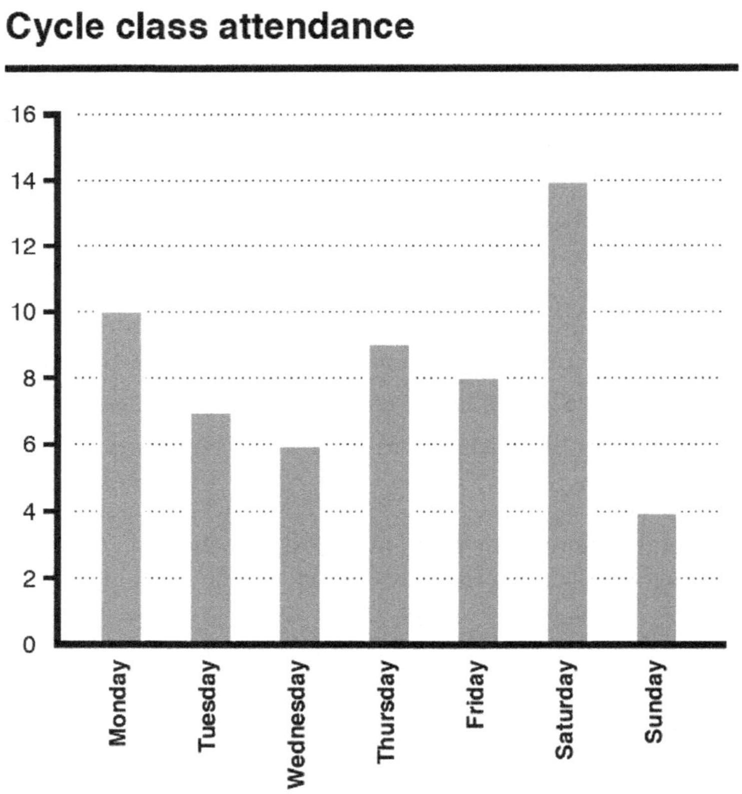

A **histogram** is a bar graph used to group data into "bins" that cover a range on the horizontal, or x-axis. Histograms consist of rectangles whose heights are equal to the frequency of a specific category. The horizontal axis represents the specific categories. Because they cover a range of data, these bins have no gaps between bars, unlike the bar graph above. In a histogram showing the heights of adult golden retrievers, the bottom axis would be groups of heights, and the y-axis would be the number of dogs in each range. Evaluating this histogram would show the height of most golden retrievers as falling within a certain range. It also provides information to find the average height and range for how tall golden retrievers may grow.

The following is a histogram that represents exam grades in a given class. The horizontal axis represents ranges of the number of points scored, and the vertical axis represents the number of students. For example, approximately 33 students scored in the 60 to 70 range.

Results of the exam

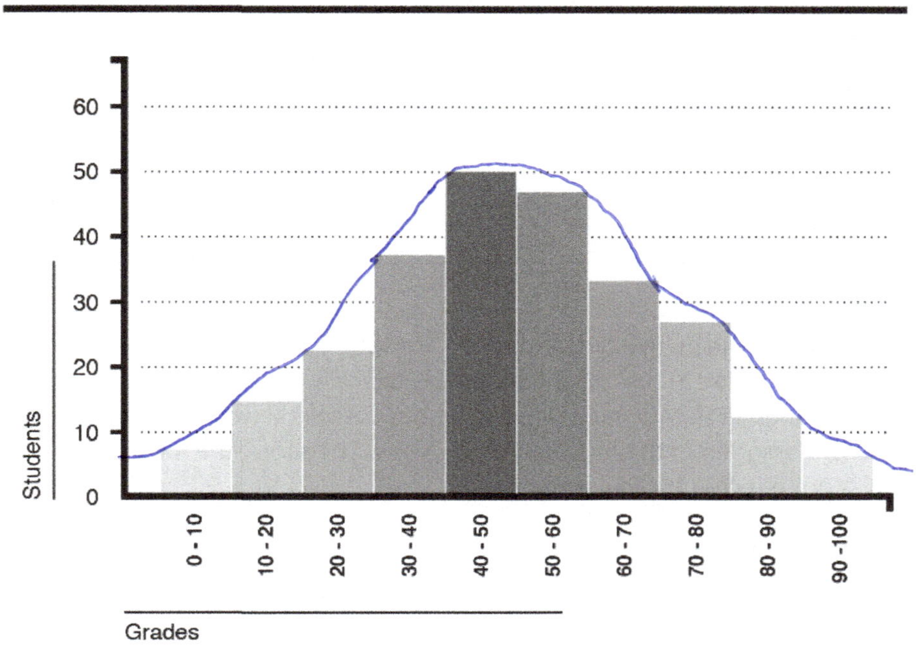

Measures of central tendency can be discussed using a histogram. If the points scored were shown with individual rectangles, the tallest rectangle would represent the mode. A bimodal set of data would have two peaks of equal height. Histograms can be classified as having data **skewed to the left, skewed to the right**, or **normally distributed**, which is also known as **bell-shaped**.

Center--mean, median
Unusal (olters)
Shape
Spread-range, standard deviation
Shape
--> approx. normal
--> skewed right
--> skewed left

These three classifications can be seen in the following chart:

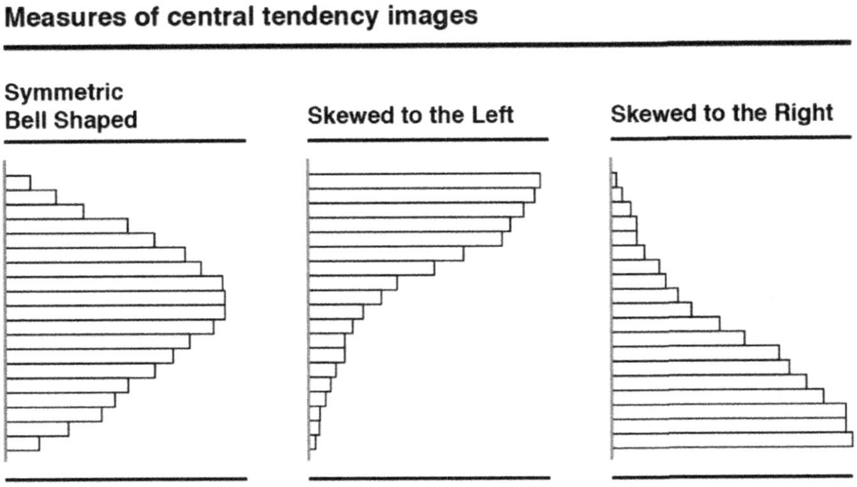

When the data is normal, the mean, median, and mode are very similar because they all represent the most typical value in the data set. In this case, the mean is typically considered the best measure of central tendency because it includes all data points. However, if the data is skewed, the mean becomes less meaningful because it is dragged in the direction of the skew. Therefore, the median becomes the best measure because it is not affected by any outliers.

The measures of central tendency and the range may also be found by evaluating information on a line graph.

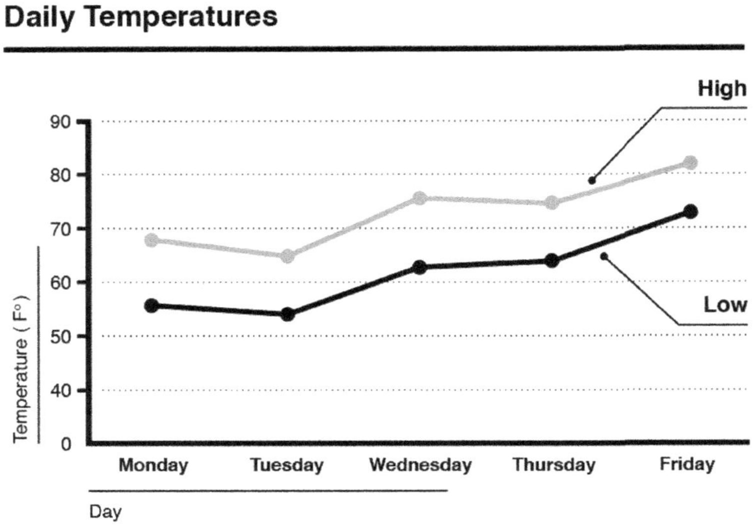

In the line graph above that shows the daily high and low temperatures, the average high temperature can be found by gathering data from each day on the triangle line. The days' highs are 69, 65, 75, 74, and 81. To find the average, add them together to get 364, then divide by 5 (because there are 5 temperatures). The average high for the five days is 72.8. If 72.8 degrees is found on the graph, it will fall in the middle of all the values. The average low temperature can be found in the same way.

Given a set of data, the **correlation coefficient**, *r*, measures the association between all the data points. If two values are **correlated**, there is an association between them. However, correlation does not necessarily mean causation, or that one value causes the other. There is a common mistake made that assumes correlation implies causation. Average daily temperature and number of sunbathers are both correlated and have causation. If the temperature increases, that change in weather causes more people to want to catch some rays. However, wearing plus-size clothing and having heart disease are two variables that are correlated but do not have causation. The larger someone is, the more likely he or she is to have heart disease. However, being overweight does not cause someone to have the disease.

The value of the correlation coefficient is between −1 and 1, where −1 represents a perfect negative linear relationship, 0 represents no relationship between the two data sets, and 1 represents a perfect positive linear relationship. A negative linear relationship means that as x-values increase, y-values decrease. A positive linear relationship means that as x-values increase, y-values increase. The formula for computing the correlation coefficient is:

$$r = \frac{n\sum xy - (\sum x)(\sum y)}{\sqrt{n(\sum x^2) - (\sum x)^2}\sqrt{n(\sum y^2) - (y)^2}}$$

where n is the number of data points. The closer r is to 1 or −1, the stronger the correlation. A correlation can be seen when plotting data. If the graph resembles a straight line, there is a correlation.

Solving Problems Involving Measures of Center and Range

As mentioned, a data set can be described by calculating the mean, median, and mode. These values allow the data to be described with a single value that is representative of the data set.

The most common measure of center is the **mean,** also referred to as the **average.**

To calculate the mean:

- Add all data values together
- Divide by the sample size (the number of data points in the set)

The **median** is middle data value, so that half of the data lies below this value and half lies above it.

To calculate the median:

- Order the data from least to greatest
- The point in the middle of the set is the median
- If there is an even number of data points, add the two middle points and divide by 2

The **mode** is the data value that occurs most often.

To calculate the mode:

- Order the data from least to greatest
- Find the value that occurs most often

Example: Amelia is a leading scorer on the school's basketball team. The following data set represents the number of points that Amelia has scored in each game this season. Use the mean, median, and mode to describe the data.

16, 12, 26, 14, 28, 14, 12, 15, 25

Solution:

Mean:

$$16 + 12 + 26 + 14 + 28 + 14 + 12 + 15 + 25 = 162$$

$$162 \div 9 = 18$$

Amelia averages 18 points per game.

Median:

12, 12, 14, 14, **15**, 16, 25, 26, 28

Amelia's median score is 15.

Mode:

12, 12, 14, 14, 15, 16, 25, 26, 28

The numbers 12 and 14 each occur twice, so this data set has 2 modes: 12 and 14.

The **range** is the difference between the largest and smallest values in the set. In the example above, the range is:

$$28 - 12 = 16$$

Determining How Changes in Data Affect Measures of Center or Range

An **outlier** is a data point that lies an unusual distance from other points in the data set. Removing an outlier from a data set will change the measures of central tendency. Removing a large outlier (a high number) from a data set will decrease both the mean and the median. Removing a small outlier (a number much lower than most in the data set) from a data set will increase both the mean and the median. For example, in data set {3, 6, 8, 12, 13, 14, 60}, the data point 60, is an outlier because it is unusually far from the other points. In this data set, the mean is 16.6. Notice that this mean number is even larger than all other data points in the set except for 60. Removing the outlier changes the mean to 9.3, and the median goes from 12 to 10. Removing an outlier will also decrease the range. In the data set above, the range is 57 when the outlier is included, but it decreases to 11 when the outlier is removed.

Adding an outlier to a data set will also affect the measures of central tendency. When a larger outlier is added to a data set, the mean and median increase. When a small outlier is added to a data set, the mean and median decrease. Adding an outlier to a data set will increase the range.

This does not seem to provide an appropriate measure of center when considering this data set. What will happen if that outlier is removed? Removing the extremely large data point, 60, is going to reduce the mean to 9.3. The mean decreased dramatically because 60 was much larger than any of the other data values. What would happen with an extremely low value in a data set like this one, {12, 87, 90, 95, 98, 100}? The mean of the given set is 80. When the outlier, 12, is removed, the mean should increase and fit more closely to the other data points. Removing 12 and recalculating the mean show that this is correct. After removing the outlier, the mean is 94. So, removing a large outlier will decrease the mean while removing a small outlier will increase the mean.

Statistical Measures

Statistics is the branch of mathematics that deals with the collection, organization, and analysis of data. A statistical question is one that can be answered by collecting and analyzing data. When collecting data, expect variability. For example, "How many pets does Yanni own?" is not a statistical question because it can be answered in one way. "How many pets do the people in a certain neighborhood own?" is a statistical question because, to determine this answer, one would need to collect data from each person in the neighborhood, and it is reasonable to expect the answers to vary.

Identify these as statistical or not statistical:

- How old are you?
- What is the average age of the people in your class?
- How tall are the students in Mrs. Jones' sixth grade class?
- Do you like Brussels sprouts?

The first and last questions are not statistical, but the two middle questions are.

Data collection can be done through surveys, experiments, observations, and interviews. A **census** is a type of survey that is done with a whole population. Because it can be difficult to collect data for an entire population, sometimes a **sample** is used. In this case, one would survey only a fraction of the population and make inferences about the data. Sample surveys are not as accurate as a census, but they are an easier and less expensive method of collecting data. An **experiment** is used when a researcher wants to explain how one variable causes changes in another variable. For example, if a researcher wanted to know if a particular drug affects weight loss, he or she would choose a **treatment group** that would take the drug, and another group, the **control group**, that would not take the drug. Special care must be taken when choosing these groups to ensure that bias is not a factor.

Bias occurs when an outside factor influences the outcome of the research. In observational studies, the researcher does not try to influence either variable but simply observes the behavior of the subjects. Interviews are sometimes used to collect data as well. The researcher will ask questions that focus on her area of interest in order to gain insight from the participants. When gathering data through observation or interviews, it is important that the researcher is well trained so that he or she does not influence the results and the study remains reliable. A study is reliable if it can be repeated under the same conditions and the same results are received each time.

The Random Processes Underlying Statistical Experiments
For researchers to make valid conclusions about population characteristics and parameters, the sample used to compare must be random. In a **random sample**, every member of the population must have an equal chance of being selected. In this situation, the sample is **unbiased** and is said to be a good representation of the population. If a sample is selected in an inappropriate manner, it is said to be **biased.** A sample can be biased if, for example, some subjects were more likely to be chosen than others. In order to have unbiased samples, the four main sampling methods used tend to be random, systematic, stratified, and cluster sampling.

Random sampling occurs when, given a sample size *n*, all possible samples of that size are equally likely to be chosen. Random numbers from calculators are typically used in this setting. Each member of a population is paired with a number, and then a set of random numbers is generated. Each person paired with one of those random numbers is selected. A **systematic sample** is when every fourth, seventh, tenth, etc., person from a population is selected to be in a sample. A **stratified sample** is when the population is divided into subgroups, or **strata**, using a characteristic, and then members from each stratum are randomly selected. For example, university students could be divided into age groups and then selected from each age group. Finally, a **cluster sample** is when a sample is used from an already selected group, like city block or zip code. These four methods are used most frequently because they are most likely to yield unbiased results.

Once an unbiased sample is obtained, data need to be collected. Common data collection methods include surveys with questions that are unbiased, contain clear language, avoid double negatives, and do not contain compound sentences that ask two questions at once. When formulating these questions, the simpler verbiage, the better.

Using Random Sampling to Draw Inferences About a Population

In statistics, a **population** contains all subjects being studied. For example, a population could be every student at a university or all males in the United States. A **sample** consists of a group of subjects from an entire population. A sample would be 100 students at a university or 100,000 males in the United States. **Inferential statistics** is the process of using a sample to generalize information concerning populations. **Hypothesis testing** is the actual process used when evaluating claims made about a population based on a sample.

A **statistic** is a measure obtained from a sample, and a **parameter** is a measure obtained from a population. For example, the mean SAT score of the 100 students at a university would be a statistic, and the mean SAT score of all university students would be a parameter.

The beginning stages of hypothesis testing starts with formulating a **hypothesis**, a statement made concerning a population parameter. The hypothesis may be true, or it may not be true. The test will answer that question. In each setting, there are two different types of hypotheses: the **null hypothesis**, written as H_0, and the **alternative hypothesis**, written as H_1. The null hypothesis represents verbally when there is not a difference between two parameters, and the alternative hypothesis represents verbally when there is a difference between two parameters. Consider the following experiment: A researcher wants to see if a new brand of allergy medication has any effect on drowsiness of the patients who take the medication. He wants to know if the average hours spent sleeping per day increases. The mean for the population under study is 8 hours, so $\mu = 8$. In other words, the population parameter is μ, the mean. The null hypothesis is $\mu = 8$ and the alternative hypothesis is $\mu > 8$. When using a smaller sample of a population, the null hypothesis represents the situation when the mean remains unaffected and the alternative hypothesis represents the situation when the mean increases. The chosen statistical test will apply the data from the sample to actually decide whether the null hypothesis should or should not be rejected.

Probabilistic Reasoning

Chance Processes and Probability Models

Probability describes how likely it is that an event will occur. Probabilities are always a number from zero to 1. If an event has a high likelihood of occurrence, it will have a probability close to 1. If there is only a small chance that an event will occur, the likelihood is close to zero. A fair six-sided die has one of the numbers 1, 2, 3, 4, 5, and 6 on each side. When this die is rolled there is a one in six chance that it will land on 2. This is because there are six possibilities and only one side has a 2 on it. The probability then is $\frac{1}{6}$ or 0.167. The probability of rolling an even number from this die is three in six, which is $\frac{1}{2}$ or 0.5. This is because there are three sides on the die with even numbers (2, 4, 6), and there are six possible sides. The probability of rolling a number less than 10 is 1; since every side of the die has a number less than 6, it would be impossible to roll a number 10 or higher. On the other hand, the probability of rolling a number larger than 20 is zero. There are no numbers greater than 20 on the die, so it is certain that this will not occur, thus the probability is zero.

If a teacher says that the probability of anyone passing her final exam is 0.2, is it highly likely that anyone will pass? No, the probability of anyone passing her exam is low because 0.2 is closer to zero than to 1. If another teacher is proud that the probability of students passing his class is 0.95, how likely is it that a student will pass? It is highly likely that a student will pass because the probability, 0.95, is very close to 1.

A probability experiment is a repeated action that has a specific set of possible results. The result of such an experiment is known as an **outcome,** and the set of all potential outcomes is known as the **sample space.** An **event** consists of one or more of those outcomes. For example, consider the probability experiment of tossing a coin and rolling a six-sided die. The coin has two possible outcomes—a heads or a tails—and the die has six possible outcomes—rolling each number 1–6. Therefore, the sample space has twelve possible outcomes: a heads or a tails paired with each roll of the die.

A **simple event** is an event that consists of a single outcome. For instance, selecting a queen of hearts from a standard fifty-two-card deck is a simple event; however, selecting a queen is not a simple event because there are four possibilities.

Classical, or **theoretical, probability** is when each outcome in a sample space has the same chance to occur. The probability for an event is equal to the number of outcomes in that event divided by the total number of outcomes in the sample space. For example, consider rolling a six-sided die. The probability of rolling a 2 is $\frac{1}{6}$, and the probability of rolling an even number is $\frac{3}{6}$, or $\frac{1}{2}$, because there are three even numbers on the die. This type of probability is based on what should happen in theory but not what actually happens in real life.

Empirical probability is based on actual experiments or observations. For example, if a die is rolled eight times, and a 1 is rolled two times, the empirical probability of rolling a 1 is $\frac{2}{8} = \frac{1}{4}$, which is higher than the theoretical probability. The Law of Large Numbers states that as an experiment is completed repeatedly, the empirical probability of an event should get closer to the theoretical probability of an event.

The **addition rule** is necessary to find the probability of event *A* or event *B* occurring or both occurring at the same time. If events *A* and *B* are **mutually exclusive** or **disjoint,** which means they cannot occur at the same time:

$$P(A \text{ or } B) = P(A) + P(B)$$

If events *A* and *B* are not mutually exclusive:

$$P(A \text{ or } B) = P(A) + P(B) - P(A \text{ and } B)$$

where $P(A \text{ and } B)$ represents the probability of event *A* and *B* both occurring at the same time. An example of two events that are mutually exclusive are rolling a 6 on a die and rolling an odd number on a die.

The probability of rolling a 6 or rolling an odd number is:

$$\frac{1}{6}+\frac{3}{6}=\frac{4}{6}=\frac{2}{3}$$

Rolling a 6 and rolling an even number are not mutually exclusive because there is some overlap. The probability of rolling a 6 or rolling an even number is:

$$\frac{1}{6}+\frac{3}{6}-\frac{1}{6}=\frac{3}{6}=\frac{1}{2}$$

Conditional Probability

The **multiplication rule** is necessary when finding the probability that event A occurs in a first trial and event B occurs in a second trial, which is written as $P(A \text{ and } B)$. This rule differs if the events are independent or dependent. Two events *A* and *B* are **independent** if the occurrence of one event does not affect the probability that the other will occur. If *A* and *B* are not independent, they are **dependent**, and the outcome of the first event somehow affects the outcome of the second. If events *A* and B are independent:

$$P(A \text{ and } B) = P(A)P(B)$$

and if events *A* and *B* are dependent:

$$P(A \text{ and } B) = P(A)P(B|A)$$

where $P(B|A)$ represents the probability event *B* occurs given that event *A* has already occurred.

$P(B|A)$ represents **conditional probability**, or the probability of event *B* occurring given that event *A* has already occurred. $P(B|A)$ can be found by dividing the probability of events *A* and *B* both occurring by the probability of event *A* occurring using the formula $P(B|A) = \frac{P(A \text{ and } B)}{P(A)}$ and represents the total number of outcomes remaining for *B* to occur after *A* occurs. This formula is derived from the multiplication rule with dependent events by dividing both sides by $P(A)$. Note that $P(B|A)$ and $P(A|B)$ are not the same. The first quantity shows that event *B* has occurred after event *A*, and the second quantity shows that event *A* has occurred after event *B*. Incorrectly interchanging these ideas is known as **confusing the inverse**.

Consider the case of drawing two cards from a deck of fifty-two cards. The probability of pulling two queens would vary based on whether the initial card was placed back in the deck for the second pull. If the card is placed back in, the probability of pulling two queens is:

$$\frac{4}{52}\times\frac{4}{52}=0.00592.$$

If the card is not placed back in, the probability of pulling two queens is:

$$\frac{4}{52}\times\frac{3}{51}=0.00452$$

When the card is not placed back in, both the numerator and denominator of the second probability decrease by 1. This is due to the fact that, theoretically, there is one less queen in the deck, and there is one less total card in the deck as well.

Using Probability to Evaluate the Outcomes of Decisions

A **two-way frequency table** displays categorical data with two variables, and it highlights relationships that exist between those two variables. Such tables are used frequently to summarize survey results and are also known as **contingency tables**. Each cell shows a count pertaining to that individual variable pairing, known as a **joint frequency**, and the totals of each row and column are also in the table.

Consider the following two-way frequency table:

Distribution of the Residents of a Particular Village

	70 or older	69 or younger	Totals
Women	20	40	60
Men	5	35	40
Total	25	75	100

The table shows the breakdown of ages and sexes of 100 people in a particular village. The end of each row or column displays the number of people represented by the corresponding data, and the total number of people is shown in the bottom right corner. For example, there were 25 people aged 70 or older and 60 women in the data. The 20 in the first cell shows that out of 100 total villagers, 20 were women aged 70 or older. The 5 in the cell below shows that out of 100 total villagers, 5 were men aged 70 or older.

A two-way table can also show relative frequencies by indicating the percentages of people instead of the count. If each frequency is calculated over the entire total of 100, the first cell would be 20% or 0.2. However, the relative frequencies can also be calculated over row or column totals. If row totals were used, the first cell would be:

$$\frac{20}{60} = 0.333 = 33.3\%$$

If column totals were used, the first cell would be:

$$\frac{20}{25} = 0.8 = 80\%.$$

Such tables can be used to calculate **conditional probabilities**, which are probabilities that an event occurs, given another event. Consider a randomly-selected villager. The probability of selecting a male 70 years old or older is $\frac{5}{100} = 0.05$ because there are 5 males over the age of 70 and 100 total villagers.

Practice Test #1

1. Using the graph below, what is the mean number of visitors for the first 4 hours?

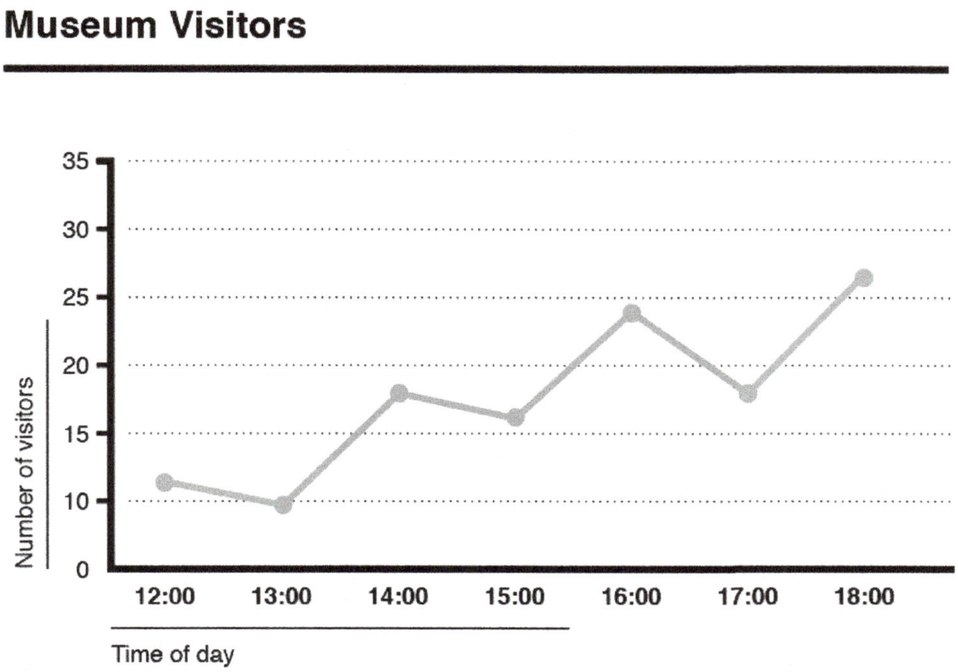

a. 12
b. 13
c. 14
d. 15

2. Is the function $g(x) = 7x^3 + 5x - 2$ odd, even, both even and odd, or neither even nor odd?
a. Odd
b. Even
c. Both
d. Neither

3. What is the missing length *x*?

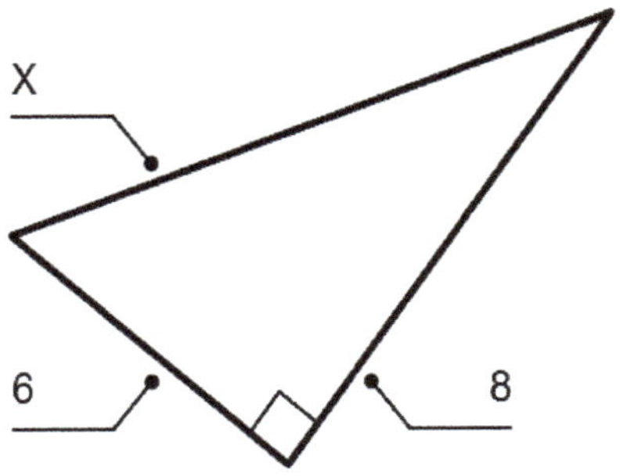

a. 6
b. 14
c. 10
d. 100

4. What is the solution to the following system of linear equations?

$$2x + y = 14$$
$$4x + 2y = -28$$

a. (0, 0)
b. (14, -28)
c. All real numbers
d. There is no solution

5. Eva Jane is practicing for an upcoming 5K run. She has recorded the following times (in minutes):

25, 18, 23, 28, 30, 22.5, 23, 33, 20

Use the above information to answer the next three questions to the closest minute. What is Eva Jane's mean time?

a. 26 minutes
b. 19 minutes
c. 25 minutes
d. 23 minutes

6. What is the mode of Eva Jane's time?

a. 16 minutes
b. 20 minutes
c. 23 minutes
d. 33 minutes

7. What is Eva Jane's median time?

a. 23 minutes
b. 17 minutes
c. 28 minutes
d. 19 minutes

8. What is the equivalent exponential form of the following logarithmic function?

$$f(x) = \log_5(x + 3)$$

a. $3^y = x$
b. $5^y = x$
c. $5^{(x+3)} = y$
d. $5^y = x + 3$

9. What is the area of the shaded region?

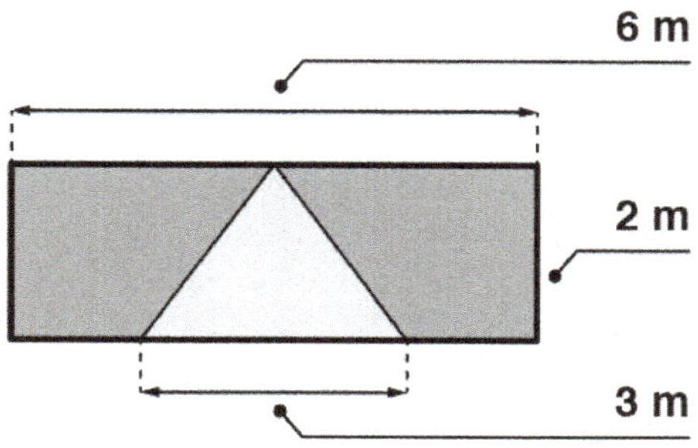

a. 9 m²
b. 12 m²
c. 6 m²
d. 8 m²

10. What is the domain of the following rational function?

$$f(x) = \frac{x^3 + 2x + 1}{2 - x}$$

a. $(-\infty, -2) \cup (-2, \infty)$
b. $(-\infty, 2) \cup (2, \infty)$
c. $(2, \infty)$
d. $(-2, \infty)$

11. What is the equation of a straight line through (0, 3) and (4, 8).

a. $y = \frac{5}{3}x + 3$
b. $y = 5x + 3$
c. $y = \frac{5}{4}x + 3$
d. $y = 4x + 3$

12. Carla is starting a cake-decorating business and wants to know how long it will take her to start making a profit. She knows the original investment is $100. After that investment, she can begin making cakes and selling them for $20 each. How many cakes will she need to sell to break even on her investment?

a. 5
b. 100
c. 10
d. 2

13. Jimmy is designing a play parachute for his PE class. The equation for the parachute to be graphed on a coordinate plane is $(x-3)^2+(y+1)^2=16$. If modeled on a graph, what point would be the center of the parachute, and what would be the radius?

a. (3, -1) and r = 16
b. (3, -1) and r = 4
c. (-3, 1) and r = 4
d. (-3, 1) and r=16

14. Solve the inequality: $|3-x|>7$.

a. $x<-4\ and\ x>-10$
b. $x>4\ and\ x<10$
c. $x<4\ and\ x>10$
d. $x>-4\ and\ x>-10$

15. What is the domain of the function: $f(x)=\frac{x+4}{x^2-1}$?

a. $x\neq-1$
b. All real numbers
c. $x\neq1$
d. $x\neq1,-1$

16. Solve: $24=e^{x-3}$.

a. 3
b. $\ln(24)+3$
c. $\ln(24)-3$
d. -3

17. What is vertex of the function $f(x)=x^2-4x+10$?

a. $(2,6)$
b. $(2,22)$
c. $(-2,22)$
d. $(-2,8)$

18. Simplify $\sqrt{16x^3y^6z^5}$.

a. $16xy^3z$
b. $4xy^3z^2\sqrt{xz}$
c. $4x^2y^3z$
d. $4xyz\sqrt{z}$

19. What will be the resulting equation if the graph of y = x^2 is shifted to the right 2 units and up 5 units?

a. $y=(x+2)^2+5$
b. $y=(x-5)^2+2$
c. $y=(x-2)^2-5$
d. $y=(x-2)^2+5$

20. If $f(x) = \sqrt{x}$ and $g(x) = -\sqrt{x+1}$, describe the translation of the graph from $f(x)$ to $g(x)$.

a. Reflected across the x-axis and translated 1 unit to the right
b. Reflected across the x-axis and translated 1 unit to the left
c. Reflected across the y-axis and translated 1 unit to the right
d. Reflected across the y-axis and translated 1 unit to the left

Answer Explanations #1

1. C: The mean for the number of visitors during the first 4 hours is 14. The mean is found by calculating the average for the four hours. Adding up the total number of visitors during those hours gives $12 + 10 + 18 + 16 = 56$. Then,

$$56 \div 4 = 14$$

2. D: To determine whether a function is even or odd, plug $-x$ into the function. If the result is $f(x)$ the function is even, and if the result is – $f(x)$ the function is odd.

$$g(-x) = 7(-x)^3 + 5(-x) - 2$$

$$-7x^3 - 5x - 2$$

This function is neither $f(x)$ nor $-f(x)$, so the given function is neither even nor odd.

3. C: The Pythagorean Theorem can be used to find the missing length x because it is a right triangle. The theorem states that $6^2 + 8^2 = x^2$, which simplifies into $100 = x^2$. Taking the positive square root of both sides results in the missing value $x = 10$.

4. D: This system can be solved using the method of substitution. Solving the first equation for y results in $y = 14 - 2x$. Plugging this into the second equation gives $4x + 2(14 - 2x) = -28$, which simplifies to $28 = -28$, an untrue statement. Therefore, this system has no solution because no x value will satisfy the system.

5. C: The mean is found by adding all the times together and dividing by the number of times recorded.

$$25 + 18 + 23 + 28 + 30 + 22.5 + 23 + 33 + 20 = 222.5$$

$$\frac{222.5}{9} = 24.722$$

Rounding to the nearest minute, the mean is 25 minutes.

6. C: The mode is the time from the data set that occurs most often. The number 23 occurs twice in the data set, while all others occur only once, so the mode is 23 minutes.

7. A: To find the median of a data set, you must first list the numbers from smallest to largest, and then find the number in the middle. If there are two numbers in the middle, add the two numbers in the middle together and divide by 2. Putting this list in order from smallest to greatest yields 18, 20, 22.5, 23, 23, 25, 28, 30, and 33, where 23 is the middle number, so 23 minutes is the median.

8. D: Given a logarithmic function $f(x) = \log_b x$, its base is b and it can be written as $y = \log_b x$ with equivalent exponential equation $b^y = x$. In this problem, the base is 5, so its equivalent exponential equation is

$$5^y = x + 3$$

9. A: The area of the shaded region is calculated in a few steps. First, the area of the rectangle is found using the formula:

$$A = length \times width = 6 \text{ m} \times 2 \text{ m} = 12 \text{ m}^2$$

Second, the area of the triangle is found using the formula:

$$A = \frac{1}{2} \times base \times height = \frac{1}{2} \times 3 \text{ m} \times 2 \text{ m} = 3 \text{ m}^2$$

The last step is to take the rectangle area and subtract the triangle area. The area of the shaded region is:

$$A = 12 \text{ m}^2 - 3 \text{ m}^2 = 9 \text{ m}^2$$

10. B: Given a rational function, the expression in the denominator can never be equal to 0. To find the domain, set the denominator equal to 0 and solve for *x*. This results in $2 - x = 0$, and its solution is $x = 2$. This value needs to be excluded from the set of all real numbers, and therefore the domain written in interval notation is $(-\infty, 2) \cup (2, \infty)$.

11. C: One form for an equation of a straight line is $y = mx + b$, where the m is the slope and the b is the *y*-intercept. To first solve for slope, the change in *y* over the change in x, the following equation is used

$$\frac{8-3}{4-0} = \frac{5}{4}$$

When this slope is substituted into the equation and one other point is used, the value for the *y*-intercept can be found. The equation becomes:

$$8 = \frac{5}{4}(4) + b$$

Solving for b yields a value of 3, so the linear equation becomes:

$$y = \frac{5}{4}x + 3$$

12. A: The equation used to model this situation is $y = 20x - 100$, where 20 is price of each cake and 100 is the original investment. The value of x is the number of cakes and *y* is the money she makes. If this line is plotted on the graph, the x-intercept will be the number of cakes she needs to make in order to recoup her investment and break even. The x-intercept occurs when the *y*-value is zero. For this equation, setting $y = 0$ and solving for x gives a value of 5 cakes. 100 cakes would yield a profit of $1,900, ten cakes would yield a profit of $100, and 2 cakes would still leave her in the negative (-$10).

13. B: The general form for the equation of a circle is:

$$(x-h)^2 + (y-k)^2 = r^2$$

The radius of the circle is the value of r and the center of the circle has coordinates (h, k). For the given equation $(x-3)^2 + (y+1)^2 = 16$, the value for the radius is $\sqrt{16} = 4 = r$. The center of the circle has coordinates $(3, -1)$.

14. A: Solving inequalities that contain absolute values requires writing two different inequalities. For the given inequality, the two new ones are $3 - x > 7$ and $3 - x < -7$. These account for the absolute value part of the inequality. When these two are solved, the two inequalities that result from it are $x < -4$ and $x > -10$. If the two inequalities are not separated and solved in the beginning, then the answer may only contain one inequality and leave out part of the answer.

15. D: The domain of a function is the possible values of x for that equation. For the given function $f(x) = \frac{x+4}{x^2-1}$, all values of x are possible except for $x = -1$ and $x = 1$. These two values cannot exist for this function because they result in a denominator that is equal to zero. These points are either holes or asymptotes for this function. The answers listed in Choices *A* and *C* are both correct, but they only contain one answer, not both negative and positive one.

16. B: Solving this exponential equation requires the use of the natural log, ln. This is the opposite of raising e to a power. Once the natural log of both sides is taken, the equation becomes $\ln(24) = x - 3$. Then, it can be solved for x, which yields the equation:

$$x = \ln(24) + 3$$

17. A: The vertex of a quadratic function can be found by the equation $x = \frac{-b}{2a}$, using the standard form for a quadratic:

$$y = ax^2 + bx + c$$

For this function, the x-coordinate for the vertex is:

$$x = \frac{-(-4)}{2(1)} = \frac{4}{2} = 2$$

Then, the *y*-value can be found by substituting this value (x = 2) into the given function:

$$y = x^2 - 4x + 10$$

$$2^2 - 4(2) + 10 = 6$$

The vertex for the quadratic equation is $(2, 6)$.

18. B: The given radical is a square root. The values under the radical can be taken out if they are perfect squares. Identifying perfect squares and removing them gives the radical:

$$4xy^3z^2\sqrt{xz}$$

If the incorrect number of variables are taken out of the radical, then the final simplified radical will be incorrect.

19. D: The standard equation for a quadratic equation is:

$$y = a(x - h)^2 + k$$

The vertex is given by (h, k). The vertex for the parent function, y = x^2, is at (0,0). A shift of 2 units to the right and 5 units up will result in a vertex of (2, 5). Therefore, the equation for the transformed graph is:

$$y = (x - 2)^2 + 5$$

20. B: If a point, (x, y) is found on the graph of $f(x)$, then the point (x, -y) is found on the graph of $-f(x)$. Note that the outputs become negative. Therefore, the graph of $-f(x)$ is a reflection of $f(x)$ across the x-axis. In an equation of the form $f(x) = \sqrt{x - a} + b$, the graph of the parent function, $f(x) = \sqrt{x}$, is translated *a* units to the right and *b* units up. For the function given by $g(x) = -\sqrt{x + 1}$, the graph of $f(x) = \sqrt{x}$ is reflected across the x-axis and translated left one unit.

Practice Test #2

1. Use the graph below entitled "Projected Temperatures for Tomorrow's Winter Storm" to answer the question.

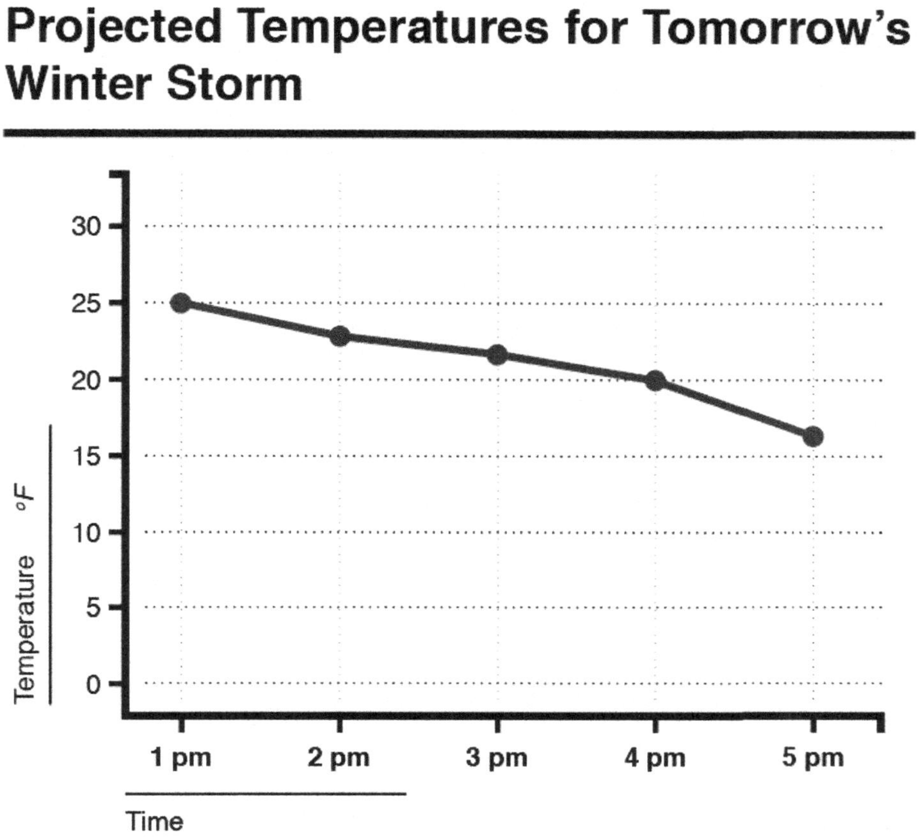

What is the expected temperature at 3:00 p.m.?

a. 25 degrees
b. 22 degrees
c. 20 degrees
d. 16 degrees

2. The function $f(x) = 3.1x + 240$ models the total U.S. population, in millions, x years after the year 1980. Use this function to answer the following question: What is the total U.S. population in 2011? Round to the nearest million.

a. 336 people
b. 336 million people
c. 6,474 people
d. 647 million people

3. What are the zeros of the following quadratic function?

$$f(x) = 2x^2 - 12x + 16$$

a. $x = 2$ and $x = 4$
b. $x = 8$ and $x = 2$
c. $x = 2$ and $x = 0$
d. $x = 0$ and $x = 4$

4. What is the volume of the cylinder below?

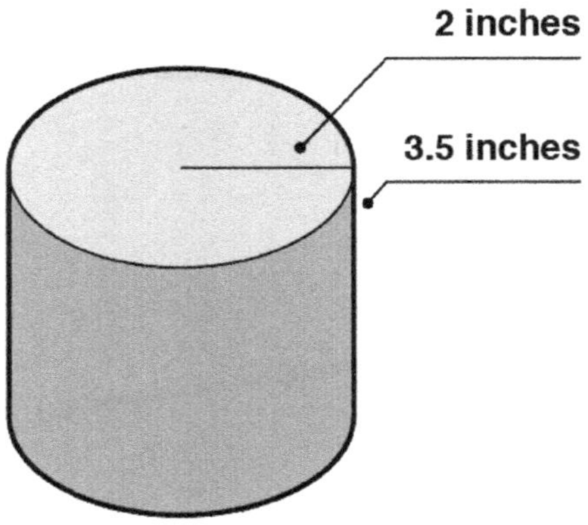

a. 18.84 in^3
b. 45.00 in^3
c. 70.43 in^3
d. 43.96 in^3

5. Given the linear function $g(x) = \frac{1}{4}x - 2$, which domain value corresponds to a range value of $\frac{1}{8}$?

a. $\frac{17}{2}$
b. $-\frac{63}{32}$
c. 0
d. $\frac{2}{17}$

6. What are the zeros of the following cubic function?

$$g(x) = x^3 - 2x^2 - 9x + 18$$

a. 2, 3
b. 2, 3, -2
c. 2, 3, -3,
d. 2, -2

7. The following set represents the test scores from a university class: {35, 79, 80, 87, 87, 90, 92, 95, 95, 98, 99}. If the outlier is removed from this set, which of the following is TRUE?

a. The mean and the median will decrease.
b. The mean and the median will increase.
c. The mean and the mode will increase.
d. The mean and the mode will decrease.

8. What is the equation of the line that passes through the two points (-3, 7) and (-1, -5)?

a. $y = 6x + 11$
b. $y = 6x$
c. $y = -6x - 11$
d. $y = -6x$

9. Which of the following is the equation of a vertical line that runs through the point (1, 4)?

a. $x = 1$
b. $y = 1$
c. $x = 4$
d. $y = 4$

10. What is the area of the following figure?

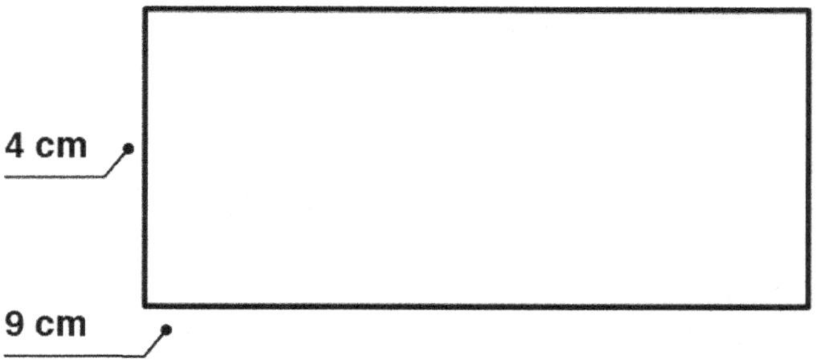

a. 26 cm
b. 36 cm
c. 13 cm^2
d. 36 cm^2

11. Which equation correctly shows how to find the surface area of a cylinder?

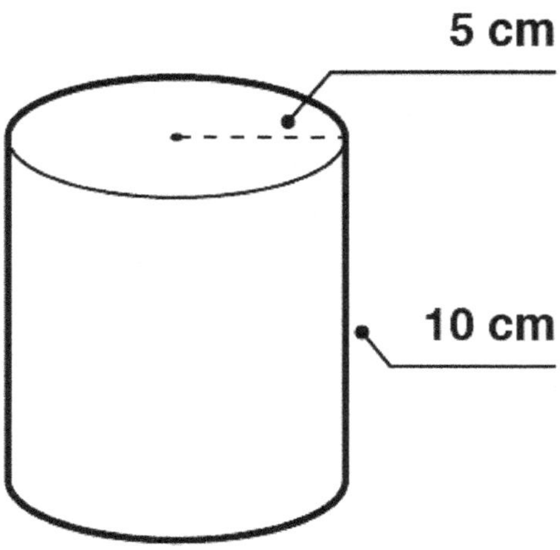

a. $SA = 2\pi \times 5 \times 10 + 2(\pi 5^2)$
b. $SA = 5 \times 2\pi \times 5$
c. $SA = 2\pi 5^2$
d. $SA = 2\pi \times 10 + \pi 5^2$

12. What is the domain of $f(x) = 4x^2 + 2x - 1$?
   a. $(0, \infty)$
   b. $(-\infty, 0)$
   c. $(-\infty, \infty)$
   d. $(-1, 4)$

13. What is the range of the polynomial function $f(x) = 2x^2 + 5$?
   a. $(-\infty, \infty)$
   b. $(2, \infty)$
   c. $(0, \infty)$
   d. $[5, \infty)$

14. For which two values of x are the following functions equal?

$$f(x) = 4x + 4$$
$$g(x) = x^2 + 3x + 2$$

   a. 1, 0
   b. -2, -1
   c. -1, 2
   d. 1, 2

15. The population of coyotes in the local national forest has been declining since 2000. The population can be modeled by the function $y = -(x - 2)^2 + 1600$, where y represents number of coyotes and x represents the number of years past 2000. When will there be no more coyotes?
   a. 2020
   b. 2040
   c. 2012
   d. 2042

16. What is the mode for the grades shown in the chart below?

Science Grades	
Jerry	65
Bill	95
Anna	80
Beth	95
Sara	85
Ben	72
Jordan	98

   a. 65
   b. 33
   c. 95
   d. 90

17. A study of adult drivers finds that it is likely that an adult driver wears his seatbelt. Which of the following could be the probability that an adult driver wears his seat belt?

a. 0.90
b. 0.05
c. 0.25
d. 0

18. Which of the following is perpendicular to the line $4x + 7y = 23$?

a. $y = -\frac{4}{7}x + 23$
b. $y = \frac{7}{4}x - 12$
c. $4x + 7y = 14$
d. $y = -\frac{7}{4}x + 11$

19. What is the solution to the following system of equations?

$$2x - y = 6$$
$$y = 8x$$

a. (1, 8)
b. (-1, 8)
c. (-1, -8)
d. There is no solution.

20. Which of the statements below is a statistical question?

a. What was your grade on the last test?
b. What were the grades of the students in your class on the last test?
c. What kind of car do you drive?
d. What was Sam's time in the marathon?

Answer Explanations #2

1. B: Look on the horizontal axis to find 3:00 p.m. Move up from 3:00 p.m. to reach the dot on the graph. Move horizontally to the left to the horizontal axis to between 20 and 25; the best answer choice is 22. The answer of 25 is too high above the projected time on the graph, and the answers of 20 and 16 degrees are too low.

2. B: The variable x represents the number of years after 1980. The year 2011 was 31 years after 1980, so plug 31 into the function to obtain:

$$f(31) = 3.1 \times 31 + 240 = 336.1$$

This value rounds to 336 and represents 336 million people.

3. A: The zeros of a polynomial function are the x-values where the graph crosses the x-axis, or where $y = 0$. Therefore, set $y = 0$ and solve the polynomial equation. This quadratic can be solved using factoring, as follows:

$$0 = 2x^2 - 12x + 16$$

$$2(x^2 - 6x + 8)$$

$$2(x - 4)(x - 2)$$

Setting both factors equal to 0 results in the two solutions $x = 4$ and $x = 2$, which are the zeros of the original function.

4. D: The volume for a cylinder is found by using the formula:

$$V = \pi r^2 h = \pi (2 \text{ in})^2 \times 3.5 \text{ in} = 43.96 \text{ in}^3$$

5. A: The range value is given, and this is the output of the function. Therefore, the function must be set equal to $\frac{1}{8}$ and solved for x. Thus, $\frac{1}{8} = \frac{1}{4}x - 2$ needs to be solved. The fractions can be cleared by multiplying by the LCD 8. This results in $1 = 2x - 16$. Add 16 to both sides and divide by 2 to obtain

$$x = \frac{17}{2}$$

6. C: To find the zeros, set the function equal to 0 and factor the polynomial. Because there are four terms, it should be factored by grouping. Factor a common factor out of the first set of two terms, and then find a shared binomial factor in the second set of two terms. This results in:

$$x^2(x - 2) - 9(x - 2) = 0$$

The binomial can then be factored out of each set to get:

$$(x^2 - 9)(x - 2) = 0$$

This can be factored further as:

$$(x + 3)(x - 3)(x - 2) = 0$$

Setting each factor equal to zero and solving results in the three zeros -3, 3, and 2.

7. B: The outlier is 35. When a small outlier is removed from a data set, the mean and the median increase. The first step in this process is to identify the outlier, which is the number that lies away from the given set. Once the outlier is identified, the mean and median can be recalculated. The mean will be affected because it averages all of the numbers. The median will be affected because it finds the middle number, which is subject to change because a number is lost. The mode will most likely not change because it is the number that occurs the most, which will not be the outlier if there is only one outlier.

8. C: First, the slope of the line must be found. This is equal to the change in y over the change in x, given the two points. Therefore, the slope is -6. The slope and one of the points are then plugged into the point-slope form of a line:

$$y - y_1 = m(x - x_1)$$

This results in:

$$y - 7 = -6(x + 3)$$

The -6 is distributed and the equation is solved for y to obtain $y = -6x - 11$.

9. A: A vertical line has the same x value for any point on the line. Other points on the line would be (1, 3), (1, 5), (1, 9), etc. Mathematically, this is written as $x = 1$. A vertical line is always of the form $x = a$ for some constant a.

10. D: The area for a rectangle is found by multiplying the length by the width. The area is also measured in square units, so the correct answer is Choice *D*. The answer of 26 is the perimeter. The answer of 13 is found by adding the two dimensions instead of multiplying.

11. A: The surface area for a cylinder is the sum of the two circle bases and the rectangle formed on the side. This is easily seen in the net of a cylinder.

The Net of a Cylinder

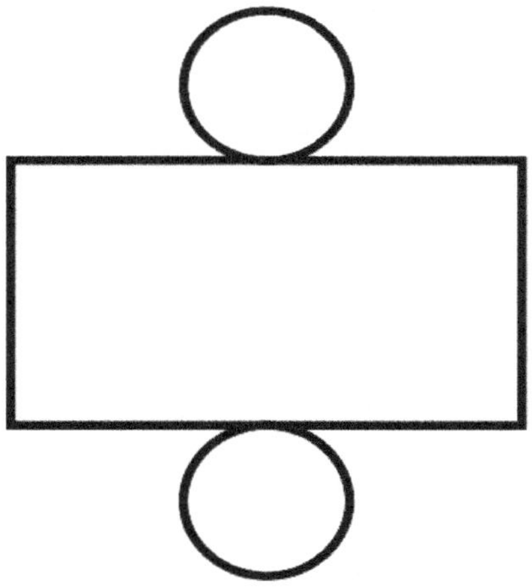

The area of a circle is found by multiplying π times the radius squared. The rectangle's area is found by multiplying the circumference of the circle by the height. The equation $SA = 2\pi \times 5 \times 10 + 2(\pi 5^2)$

shows the area of the rectangle as $2\pi \times 5 \times 10$, which yields 314. The area of the bases is found by $\pi 5^2$, which yields 78.5, then multiplied by 2 for the two bases.

12. C: The function given is a polynomial function. Anything can be plugged into a polynomial function to get an output. Therefore, its domain is all real numbers, which is expressed in interval notation as $(-\infty, \infty)$.

13. D: This is a parabola that opens up, as the coefficient on the x^2 term is positive. The smallest number in its range occurs when plugging 0 into the function $f(0) = 5$. Any other output is a number larger than 5, even when a positive number is plugged in. When a negative number gets plugged into the function, the output is positive, and same with a positive number. Therefore, the domain is written as $[5, \infty)$ in interval notation.

14. C: First, set the functions equal to one another, resulting in:

$$x^2 + 3x + 2 = 4x + 4$$

This is a quadratic equation, so the best way to find the answer is to convert it to standard form by subtracting $4x + 4$ from both sides of the equation, setting it equal to 0:

$$x^2 - x - 2 = 0$$

Then factor the equation:

$$(x - 2)(x + 1) = 0$$

Setting both factors equal to zero results in $x = 2$ and $x = -1$.

15. D: There will be no more coyotes when the population is 0, so set y equal to 0 and solve the quadratic equation:

$$0 = -(x - 2)^2 + 1600$$

Subtract 1600 from both sides and divide by -1. This results in:

$$1600 = (x - 2)^2$$

Then, take the square root of both sides. This process results in the following equation:

$$\pm 40 = x - 2$$

Adding 2 to both sides results in two solutions: $x = 42$ and $x = -38$. Because the problem involves years after 2000, the only solution that makes sense is 42. Add 42 to 2000; therefore, in 2042 there will be no more coyotes.

16. C: The mode for a set of data is the value that occurs the most. The grade that appears the most is 95. It's the only value that repeats in the set.

17. A: The probability of 0.9 is closer to 1 than any of the other answers. The closer a probability is to 1, the greater the likelihood that the event will occur. The probability of 0.05 shows that it is very unlikely that an adult driver will wear their seatbelt because it is close to zero. A zero probability means that it will not occur. The probability of 0.25 is closer to zero than to one, so it shows that it is unlikely an adult will wear their seatbelt.

18. B: The slopes of perpendicular lines are negative reciprocals, meaning their product is equal to -1. The slope of the line given needs to be found. Its equivalent form in slope-intercept form is $y = -\frac{4}{7}x + \frac{23}{7}$, so its slope is $-\frac{4}{7}$. The negative reciprocal of this number is $\frac{7}{4}$. The only line in the options given with this same slope is:

$$y = \frac{7}{4}x - 12$$

19. C: This system can be solved using substitution. Plug the second equation in for y in the first equation to obtain $2x - 8x = 6$, which simplifies to $-6x = 6$. Divide both sides by 6 to get $x = -1$, which is then back-substituted into either original equation to obtain $y = -8$.

20. B: This is a statistical question because to determine this answer one would need to collect data from each person in the class and it is expected that the answers would vary. The other answers do not require data to be collected from multiple sources; therefore, the answers will not vary.

Practice Test #3

1. Given the function $f(x) = 4x - 2$, what is the correct form of the simplified difference quotient $\frac{f(x+h)-f(x)}{h}$?

   a. $4x - 1$
   b. $4x$
   c. 4
   d. $4x + h$

2. How many possible positive zeros does the polynomial function $f(x) = x^4 - 3x^3 + 2x + x - 3$ have?

   a. 4
   b. 3
   c. 2
   d. 1

3. The sides of a triangle have the following lengths: 4 inches, 4 inches, and 7.5 inches. If the smallest angle within the triangle has a measurement of 20°, what is the measure of the largest angle?

   a. 100°
   b. 140°
   c. 120°
   d. 160°

4. On the first four tests this semester, a student received the following scores out of 100: 74, 76, 82, and 84. The student must earn at least what score on the fifth test to receive a B in the class? Assume that the final test is also out of 100 points and that to receive a B in the class, he must have at least an 80% average.

   a. 80
   b. 84
   c. 82
   d. 78

5. Evaluate the expression: $5b^2 - 3ab + 10a$, when $a = 3$ and $b = -1$.

   a. 84
   b. 26
   c. 44
   d. 30

6. What is the volume of the given figure?

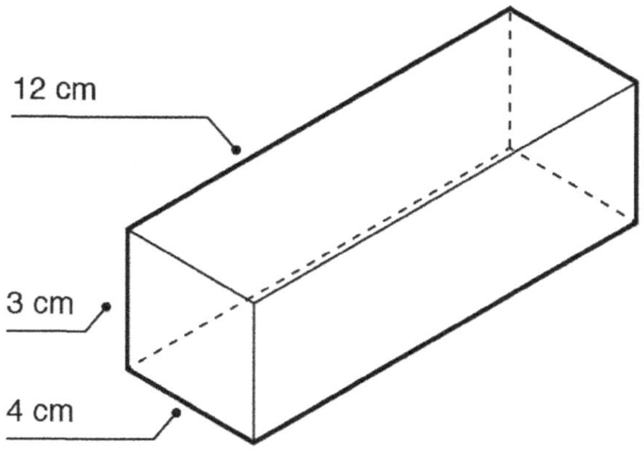

a. 36 cm^2
b. 144 cm^3
c. 72 cm^3
d. 36 cm^3

7. Solve: $5(x-2)-3x=20+x$.

a. 22
b. 15
c. 30
d. 10

8. What is the perimeter of the following figure?

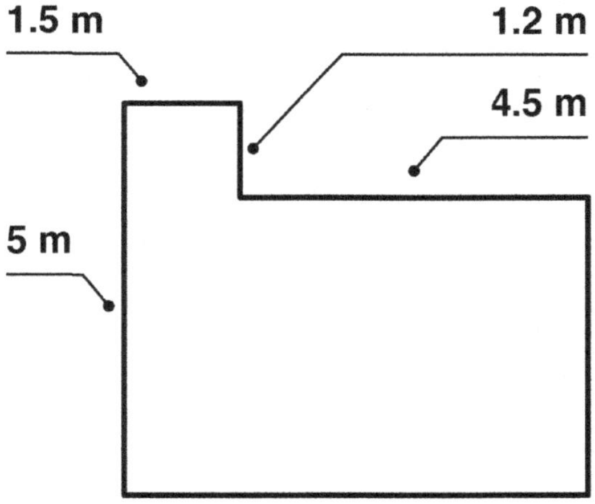

a. 13.4 m
b. 22 m
c. 12.2 m
d. 22.5 m

9. Simplify: $\frac{x^3+4x^2y-5xy^2}{x^2-xy}$.

a. $x - y$
b. $x^2 + 5xy$
c. $x + 5y$
d. $x^2 - xy$

10. A local candy store reports that of the 100 customers that bought suckers, 35 of them bought cherry. What is the probability of selecting 2 customers simultaneously at random that both purchased a cherry sucker?

a. $\frac{119}{990}$
b. $\frac{35}{100}$
c. $\frac{49}{400}$
d. $\frac{69}{99}$

11. Erin and Katie work at the same ice cream shop. Together, they always work less than 21 hours a week. In a week, if Katie worked two times as many hours as Erin, how many hours did Erin work?

a. Less than 7 hours
b. Less than or equal to 7 hours
c. More than 7 hours
d. Less than 8 hours

12. The table below shows tickets purchased during the week for entry to the local zoo. What is the mean of adult tickets sold for the week?

Day of the Week	Age	Tickets Sold
Monday	Adult	22
Monday	Child	30
Tuesday	Adult	16
Tuesday	Child	15
Wednesday	Adult	24
Wednesday	Child	23
Thursday	Adult	19
Thursday	Child	26
Friday	Adult	29
Friday	Child	38

a. 24.2
b. 21
c. 22
d. 26.4

13. The table below displays the number of three-year-olds at Kids First Daycare who are potty-trained and those who still wear diapers.

	Potty-trained	**Wear diapers**	
Boys	26	22	48
Girls	34	18	52
	60	40	

What is the probability that a three-year-old girl chosen at random from the school is potty-trained?

a. 52 percent
b. 34 percent
c. 65 percent
d. 57 percent

14. Which of the following is the result when solving the equation $4(x + 5) + 6 = 2(2x + 3)$?

a. $x = 26$
b. $x = 6$
c. All real numbers
d. There is no solution

15. Two consecutive integers exist such that the sum of three times the first and two less than the second is equal to 411. What are those integers?

a. 103 and 104
b. 104 and 105
c. 102 and 103
d. 100 and 101

16. From the chart below, which two are preferred by more men than women?

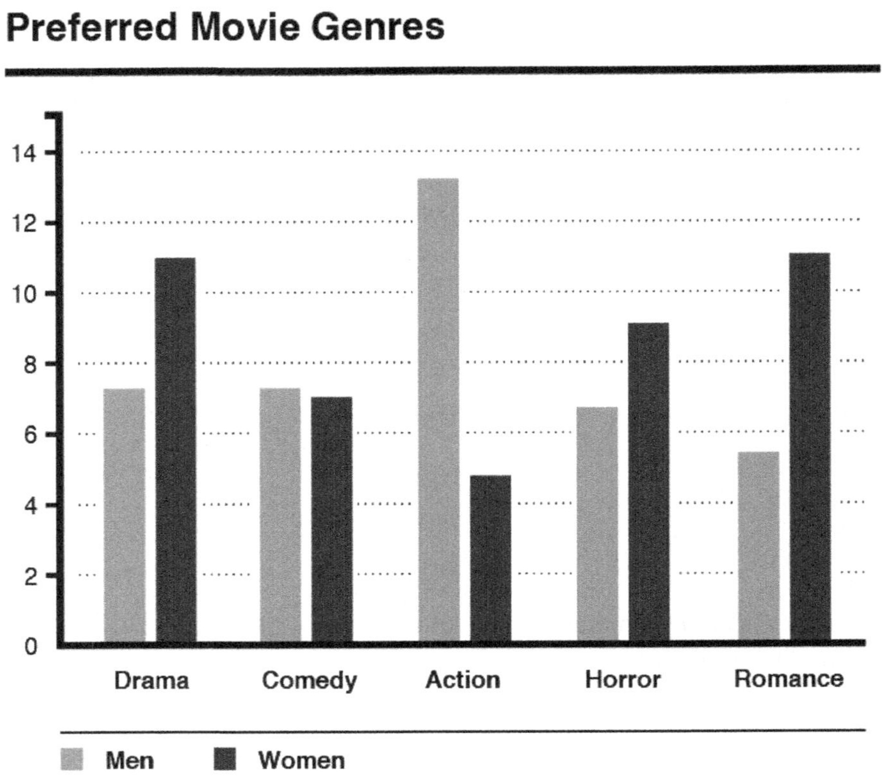

a. Comedy and Action
b. Drama and Comedy
c. Action and Horror
d. Action and Romance

17. Which type of graph best represents a continuous change over a period of time?
a. Bar graph
b. Line graph
c. Pie graph
d. Histogram

18. Which equation best represents the scatterplot below?

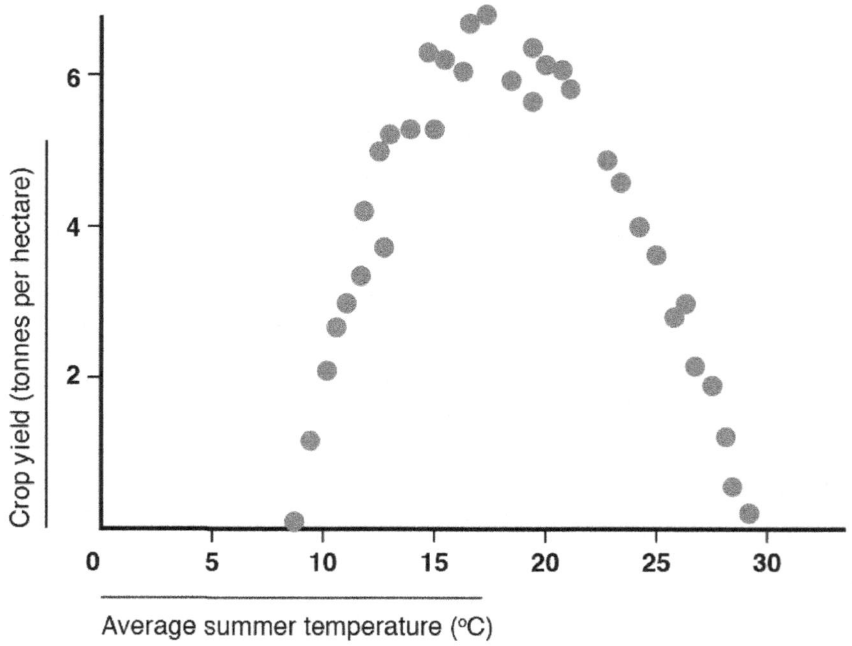

a. $y = 3x - 4$
b. $y = 2x^2 + 7x - 9$
c. $y = (3)(4^x)$
d. $y = -\frac{1}{14}x^2 + 2x - 8$

19. Find the indicated function value: $f(5)$ for $f(x) = x^2 - 2x + 1$.
a. 16
b. 1
c. 5
d. Does not exist

20. The mass of the moon is about 7.348×10^{22} kilograms and the mass of Earth is 5.972×10^{24} kilograms. How many times GREATER is Earth's mass than the moon's mass?
a. 8.127×10^1
b. 8.127
c. 812.7
d. 8.127×10^{-1}

7.3490000000

Answer Explanations #3

1. C: Plugging the function into the formula results in:

$$\frac{4(x+h)-2-(4x-2)}{h}$$

which is simplified to:

$$\frac{4x+4h-2-4x+2}{h}=\frac{4h}{h}=h$$

This value is also equal to the derivative of the given function. The derivative of a linear function is its slope.

2. B: One way to determine the number of zeros is to count the number of sign changes in coefficients in the polynomial. This results in the number of possible positive zeros. The coefficients are 1, -3, 2, 1, and -3, so the sign changes from 1 to -3, -3 to 2, and 1 to -3, a total of 3 times. Therefore, there are at most 3 positive zeros.

3. B: Because two sides are equal, this is an isosceles triangle. The smallest sides correspond to the 20° angles. The third angle has a measure of:

$$180-20-20=140°$$

4. B: Let x be equal to the fifth test score. Therefore, in order to receive, at minimum, a B in the class, the student must have:

$$\frac{74+76+82+84+x}{5}=80$$

Therefore:

$$\frac{316+x}{5}=80$$

Solving for x gives $316+x=400$, or $x=84$. Therefore, he must receive at least an 84 out of 100 on the fifth test to receive a B in the course.

5. C: Choice *C* is the correct simplification for the expression given. For the expression $5b^2-3ab+10a$, when $a=3$ and $b=-1$, the values for *a* and *b* can be substituted into the given expression. The new expression becomes:

$$5(-1)^2-3(3)(-1)+10(3)$$

Simplifying this expression using the order of operations yields a final answer of 44.

6. B: The volume of a rectangular prism is found by multiplying the length by the width by the height. This formula yields an answer of 144 cubic units. The answer must be in cubic units because volume involves all three dimensions. Each of the other answers have only two dimensions that are multiplied, and one dimension is forgotten, as in *D*, where 12 and 3 are multiplied.

7. C: Choice C simplifies the equation completely to find the value of x. The equation can be solved by working to isolate the variable. The first step in this process is to distribute the 5, which yields the equation:

$$5x - 10 - 3x = 20 + x$$

Then, the like terms can be collected on each side to form:

$$2x - 10 = 20 + x$$

Finally, the x-terms can be moved to one side and the constants can be moved to one side to find the final answer of $x = 30$.

8. B: The perimeter is found by adding the length of all the exterior sides. When the given dimensions are added, the perimeter is 22 meters. The equation to find the perimeter can be:

$$P = 5 + 1.5 + 1.2 + 4.5 + 3.8 + 6 = 22$$

The last two dimensions can be found by subtracting 1.2 from 5, and adding 1.5 and 4.5, respectively.

9. C: In order to simplify this expression, a common factor can be taken out of the numerator and denominator. The expression then becomes,

$$\frac{x(x^2 + 4xy - 5y^2)}{x(x - y)}$$

When those values for x cancel out, then the new polynomial on top can be factored into,

$$\frac{(x - y)(x + 5y)}{x - y}$$

The two equivalent binomials can be canceled, and the simplified expression becomes $x + 5y$.

10. A: The probability of choosing two customers simultaneously is the same as choosing one and then choosing a second without putting the first back into the pool of customers. This means that the probability of choosing a customer who bought cherry is $\frac{35}{100}$. Then without placing them back in the pool, it would be $\frac{34}{99}$.

So, the probability of choosing 2 customers simultaneously that both bought cherry would be:

$$\frac{35}{100} \times \frac{34}{99}$$

$$\frac{1{,}190}{9{,}900}$$

$$\frac{119}{990}$$

11. A: Let x be the unknown, the number of hours Erin can work. We know Katie works $2x$, and the sum of all hours is less than 21. Therefore, $x + 2x < 21$, which simplifies into $3x < 21$. Solving this results in the inequality $x < 7$ after dividing both sides by 3. Therefore, Erin worked less than 7 hours.

12. C: To find the mean, or average, of a set of values, add the values together and then divide by the total number of values. Each day of the week has an adult ticket amount sold that must be added together. The equation is as follows:

$$\frac{22 + 16 + 24 + 19 + 29}{5} = 22$$

13. C: The conditional frequency of a girl being potty-trained is calculated by dividing the number of potty-trained girls by the total number of girls: $34 \div 52 = 0.65$. To determine the conditional probability, multiply the conditional frequency by 100: $0.65 \times 100 = 65\%$.

14. D: The distributive property is used on both sides to obtain:

$$4x + 20 + 6 = 4x + 6$$

Then, like terms are collected on the left, resulting in:

$$4x + 26 = 4x + 6$$

Next, the addition principle is used to subtract $4x$ from both sides, and this results in the false statement $26 = 6$. Therefore, there is no solution.

15. A: First, the variables have to be defined. Let x be the first integer; therefore, $x + 1$ is the second integer. This is a two-step problem. The sum of three times the first and two less than the second is translated into the following expression:

$$3x + (x + 1 - 2)$$

This expression is set equal to 411 to obtain:

$$3x + (x + 1 - 2) = 411$$

The left-hand side is simplified to obtain:

$$4x - 1 = 411$$

The addition and multiplication properties are used to solve for x. First, add 1 to both sides and then divide both sides by 4 to obtain $x = 103$. The next consecutive integer is 104.

16. A: The chart is a bar chart showing how many men and women prefer each genre of movies. The dark gray bars represent the number of women, while the light gray bars represent the number of men. The light gray bars are higher and represent more men than women for the genres of Comedy and Action.

17. B: A line graph represents continuous change over time. The line on the graph is continuous and not broken, as on a scatter plot. A bar graph may show change but isn't necessarily continuous over time. A pie graph is better for representing percentages of a whole. Histograms are best used in grouping sets of data in bins to show the frequency of a certain variable.

18. D: The shape of the scatterplot is a parabola (U-shaped). This eliminates Choices *A* (a linear equation that produces a straight line) and *C* (an exponential equation that produces a smooth curve upward or downward). The value of a for a quadratic function in standard form ($y = ax^2 + bx + c$) indicates whether the parabola opens up (U-shaped) or opens down (upside-down U). A negative value for a produces a parabola that opens down; therefore, Choice *B* can also be eliminated.

19. A: To find a function value, plug in the number given for the variable and evaluate the expression, using the order of operations (parentheses, exponents, multiplication, division, addition, subtraction). The function given is a polynomial function:

$$f(5) = 5^2 - 2 \times 5 + 1$$

$$f(5) = 25 - 10 + 1 = 16$$

20. A: Division can be used to solve this problem. The division necessary is:

$$\frac{5.972 \times 10^{24}}{7.348 \times 10^{22}}$$

To compute this division, divide the constants first then use algebraic laws of exponents to divide the exponential expression.

This results in about 0.8127×10^2, which written in scientific notation is 8.127×10^1.

Greetings!

First, we would like to give a huge "thank you" for choosing us and this study guide for the math section of your TSI exam. We hope that it will lead you to success on this exam and for your years to come.

Our team has tried to make your preparations as thorough as possible by covering all of the topics you should be expected to know. In addition, our writers attempted to create practice questions identical to what you will see on the day of your actual test. We have also included many test-taking strategies to help you learn the material, maintain the knowledge, and take the test with confidence.

We strive for excellence in our products, and if you have any comments or concerns over the quality of something in this study guide, please send us an email so that we may improve.

As you continue forward in life, we would like to remain alongside you with other books and study guides in our library, such as:

SAT: amazon.com/dp/1628458569

ACT: amazon.com/dp/1628457597

We are continually producing and updating study guides in several different subjects. If you are looking for something in particular, all of our products are available on Amazon. You may also send us an email!

Sincerely,
APEX Test Prep
info@apexprep.com

Made in the USA
Coppell, TX
22 June 2021

57906168R00066